Library Development

Library Development

Editor

Silvius Mateus

scitus
academics

Library Development
Edited by **Silvius Mateus**

ISBN: 978-1-68117-562-1
Library of Congress Control Number: 2016942976

SCITUS Academics LLC,
www.scitusacademics.com
Box No. 4766, 616 Corporate Way,
Suite 2, Valley Cottage,
NY 10989

Notice

Printed in United States of America on Acid Free Paper

Preface

Library development is the process of meeting the information needs of the people (a service population) in a timely and economical manner using information resources locally held, as well as from other organisations. A library is a collection of sources of information and similar resources, made accessible to a defined community for reference or borrowing. It provides physical or digital access to material, and may be a physical building or room, or a virtual space, or both. A library's collection can include books, periodicals, newspapers, manuscripts, films, maps, prints, documents, microform, CDs, cassettes, videotapes, DVDs, Blu-ray Discs, e-books, audiobooks, databases, and other formats. Library development provides leadership, grants, and other assistance to libraries, archives, and literacy communities to improve services. It includes development of programs and facilities; administers funds for improvement of services; conducts on-site visits; writes proposals and collaborates with local, state and federal agencies to develop and implement programs; collects, compiles and disseminates statistics; supports training opportunities for local staff and trustees; publishes directories; newsletters, plans and brochures; and facilitates statewide planning efforts.

Library Development presents comprehensive overview to reflect the increasing integration of library, information center, records, information technology and telecommunications management, this book takes an integrated approach to managing the modern information center.

Table of Contents

Chapter 1

Designing Libraries for Research Collaboration in the Network World: An Exploratory Study

Sheila Corrall

University of Pittsburgh, USA

ABSTRACT

Advances in technology and changes in higher education and research are forcing libraries to rethink the services they offer to researchers and how they are presented. Librarians are moving into "higher end support" and adopting new service models based on "deep collaboration" with academic partners. The internal arrangements that form the design of an organization can determine whether it achieves its mission under particular conditions. The present study explores how libraries in 24 leading UK research universities are organizing resources and services to support the research enterprise. Qualitative data were collected from institutional websites and other public domain sources and were analysed using matrix techniques. The results show a trend away from integrated library and computing service organizations; variety in institutional reporting lines, but predominantly in large professional service groups; consistency in internal library groupings, but variation in portfolio and job titles; expansion of specialist positions, with new functional roles complementing traditional subject liaisons; and dedicated spaces, working groups, and integrated websites promoting boundary-spanning activities. The findings confirm and extend prior work and are being used to design a large scale international survey.

KEYWORDS

management structures, organization design, research support, university libraries

1. INTRODUCTION

Developments in digital media, network technologies, research workflows, scholarly communication and funding policies are challenging academic libraries to respond with a wider array of services and facilities for researchers (**Bourg, Coleman, & Erway, 2009**; **Housewright, Schonfeld, & Wulfson, 2013a**; **Vinopal & McCormick, 2013**). Subject/liaison librarians have traditionally provided research services as part of their academic support role, but libraries are increasingly identifying research support as a specific area requiring additional co-ordination and strategic development:

*"Open-access publishing, institutional repositories, the need to co-ordinate collection management and storage, the increasing availability of information technology (IT) tools to help researchers with their work and, not least, the Research Excellence Framework (REF) have all pointed to the need for a coherent library strategy and response" (**Bradbury & Weightman, 2010, p. 65**).*

University library support for research has emerged as a strategic issue and generated studies investigating service developments, new roles, skills requirements (**Auckland, 2012**; **Corrall, Kennan, & Afzal, 2013**; **Covert-Vail & Collard, 2012**; **Jaguszewski & Williams, 2013**). The present paper offers a different perspective on the topic by examining library engagement with research and researchers through the lens of organizational design.

1.1 Background

There is a perceived requirement for libraries to develop more specialized interventions to provide point-of-need research support and advice around the whole knowledge creation cycle: from ideas generation and project conception, through data acquisition, manipulation and interpretation, to the deposit of results, publication of findings, and assessment of impact (**Larsen & Riis, 2012**;**MacColl & Jubb, 2011**; **Vaughan et al., 2013**). Libraries are accordingly moving into areas such as funding opportunities and grant writing, ethics review, data curation and repository management, poster design and conference hosting (physical and virtual), journal and monograph publishing, bibliometric evaluation and impact assessment (**Adema & Schmidt, 2010**; **Carroll, 2011a**;**Crow et al., 2012**; **Delserone, Kelly, & Kempf, 2010**; **Drummond & Wartho, 2009**;**Furlough, 2010**; **Healy, 2010**; **Herther, 2009**; **Lyon, 2012**; **McBain, Culshaw, & Walkley Hall, 2013**; **Vaughan et al., 2013**), alongside their established roles in collection stewardship, resource procurement and information literacy. The development of support for the management of research data in line with national policy is a particular concern (**Carlson & Kneale, 2011**; **Gold, 2007**;**Lynch, 2008**; **Lyon, 2012**).

Commentators increasingly assert that libraries need to reposition from support service to professional/scholarly partner and to transform their liaison librarians into embedded experts recognized as academic associates (**Carlson & Kneale, 2011**; **Duranceau, 2008**; **Fonseca & Viator, 2009**; **Gold, 2007**; **Vandegrift & Varner, 2013**), so that they can add value through "higher end research support" (**Corrall et al., 2013, p.**

638). **Vaughan et al. (2013, p. 313)** argue that service presentation is an important factor here: "By basing the service model on the research lifecycle, the library is then poised to be a partner through the entire process, not just at the bookends of research". The proposed shift from the "service-and-support" approach to a partnership model is significant. **Posner (2013, p. 45)** argues that contemporary digital humanities projects "do not need*supporters* – they need *collaborators*", explaining that libraries need to provide both infrastructure (tools, servers, etc.) and "intellectual labor" (knowledgeable librarians). **Vandegrift and Varner (2013, pp. 69, 76)** similarly explain how partnering scholars in digital humanities should go beyond collection building to content creation, by "making 'stuff'" (e.g., web sites, digitized collections, new tools), describing the model as "deep collaboration".

Others have questioned whether libraries have sufficient expertise for such roles, particularly the technical understanding and domain knowledge needed for data management and curation in different disciplines (**Gold, 2007**; **Lynch, 2008**). However, feedback from biomedical and science librarians collaborating as partners in a large-scale cross-disciplinary multi- institutional research project reveals that personal attitudes and interpersonal skills are the key factors for successful participation in team science, with the willingness to learn and adapt more important than a deep understanding of the technology (**García-Milian et al., 2013**). Sector organizations including Research Libraries UK, the Joint Information Systems Committee, and the Association of Research Libraries (ARL), have sponsored projects examining the skillsets needed for new or extended subject liaison and specialist roles (**Auckland, 2012**; **Cox, Verbaan, & Sen, 2012**; **Jaguszewski & Williams, 2013**). Library and information schools are also responding with modules, courses, specializations, internships, and institutes to prepare new professionals and update the existing workforce (**Creamer, Morales, Kafel, Crespo, & Martin, 2012**; **Harris-Pierce & Liu, 2012**;**Lee & Tibbo, 2011**; **Stanton et al., 2011**; **Yakel, Conway, Hedstrom, & Wallace, 2011**).

The model of the librarian as a "scholar-practitioner" and active researcher is being promoted in several institutions (**Giesecke, 2010**; **McBain et al., 2013**;**Schrader, Shiri, & Williamson, 2012**), and being used to bring people with PhDs but no library degrees into research support roles in "alt/ac" (alternative to the academy) positions, particularly in relation to digital humanities developments (**Brunner, 2010**; **Giesecke, 2010**; **Nowviskie, 2013**; **Shore, 2012**). Even so, **MacColl and Jubb (2011, p. 10)** point out that despite recognizing their need for support, "researchers have little interest in the support services libraries have built for them". **Jahnke and Asher (2012, p. 16)** similarly report that "Few researchers see the library as a partner". Two large-scale surveys of academic faculty conducted in the US and UK confirm this message, with only half of the US respondents and one-third of the UK sample rating the library role in supporting their research productivity as "very important", placing this role as the least important of the six service areas specified (**Housewright et al., 2013a**; **Housewright, Schonfeld & Wulfson, 2013b**).

Some research libraries are responding to the challenge by moving beyond alignment with academic departments or schools to focus provision of in-depth information advice at the more specialized and dynamic level of research groups and projects, particularly in the biomedical and health sciences area (**Carlson & Kneale, 2011**; **Federer, 2013**; **Greyson & Surette, 2013**). Another significant trend is the creation of specialist positions dedicated to research support, either exclusively or in addition to a subject librarian role (**Bradbury & Weightman, 2010**; **Young & Lund, 2008**), including positions coordinating support for graduate students (**Covert-Vail & Collard, 2012**). The 2013 Ithaka survey of US library directors indicates that almost two-thirds of respondents in doctoral institutions expect to increase staffing for specialized faculty research support (**Long & Schonfeld, 2014**). Others have decided that traditional structures, such as the subject-specialist information consultant model, are no longer viable, and have reorganized their staffing to align with the broad institutional missions of research and teaching, or adopted business enterprise structures, with teams dedicated to service development and innovation (**Andrade & Zaghloul, 2010**;**Cotta-Schønberg, 2007**; **Covert-Vail & Collard, 2012**; **Drummond & Wartho, 2009**; **Franklin, 2009**; **Manchester University Library, 2012**; **Nutefall & Chadwell, 2012**).

1.2 Related work

Organization design has not received due attention in the debate on reshaping the library as research collaborator. Current literature on academic library structures consists largely of institutional case studies (**Andrade & Zaghloul, 2010**; **Chan & Soong, 2011**; **Franklin, 2009**; **Larsen & Riis, 2012**; **Nesdill, Love & Hunt, 2010**; **Quinlan & McHarg, 2012**; **Renfro & Neal, 2012**; **Shore, 2012**; **Warren, 2011**), with few comprehensive surveys. In North America, ARL last surveyed library organization in 1996, although the organization and staffing of specific elements of research support have been examined more recently, including digital humanities, research data management, and scholarly communication (**Bryson, Posner, St. Pierre, & Varner, 2011**; **Eustis & Kenney, 1996**; **Fearon, Gunia, Pralle, Lake & Sallans, 2013**; **Radom, Feltner-Reichert, & Stringer-Stanback, 2012**). In the UK, the convergence and merging of university libraries with information technology and other academic support services has dominated the discussion of library structures over the past two decades (**Bulpitt, 2012**; **Field, 2001**; **Hanson, 2005**; **Pugh, 1997a**).

Few writers on library organization relate their discussion to classic organizational theory, notable exceptions being **Pugh's (1997b, 2005, 2007)** efforts to establish a theoretical basis for the management of technology-driven change in libraries and **McGuigan's (2012)** essay identifying theories relevant to managing change in academic libraries. Although library case studies rarely use terminology from organization design literature, it is evident that the dominant library model of recent decades is basically the mixed structure (**Child,**

1988), which organizes technical/back-office processes (e.g., acquisitions, cataloging) and generic front-of-house services (e.g., circulation) along functional lines and specialized client services in a divisional or market-based arrangement (i.e., individuals or teams devoted to supporting particular subjects, disciplines, departments, or schools). Many libraries have overlaid an element of matrix structure, by introducing cross-cutting roles to coordinate service provision across different groups (e.g., designating senior liaison librarians as lead specialist for key areas, such as scholarly communications); an emerging variant of this model uses functional specialists who do not have liaison assignments to supplement and support their liaisons (**Jaguszewski & Williams, 2013**).

1.3 Definition of key terms

Organization design has been defined as "the allocation of resources and people to a specified mission or purpose and the structuring of these resources to achieve the mission" (**Duncan, 1979**, p. 60). **Duncan (1979, p. 60)** also asserts that "Ideally, the organization is designed to fit its *environment* and to provide the information and *coordination* needed" (emphasis added). **Stanford (2013, p. 19)** offers an alternative definition of the concept emphasizing other dimensions:

"Arranging how to do the work necessary to effectively and efficiently achieve a business purpose and strategy whilst delivering high-quality customer and employee experience. Arranging involves aligning the organization with the strategy, creating coherent designs, while building trust among key stakeholders."

Design is thus conceived as the broader higher-level concept, with *structure*forming an important part of the design. However, the terms are often used synonymously, within both management and library literature. Another important point to note here is that "resources" in this context can include "tangible resources such as money and equipment and intangible resources such as knowledge" (**Miles, Snow, Fjeldstad, Miles, & Lettl, 2010, p. 93**).

1.4 Research question, aims and objectives

The central research question for the study is: *How are academic research libraries and their institutions organizing resources and services to support the research enterprise?*

The overall aims of the research are:

- To review organizational designs of research libraries and evaluate their fitness for purpose in the context of researcher support needs.
- The specific objectives of the project are:
- To review trends and developments in library support and services for research and researchers
- To examine design dimensions and structural features of a sample of research library organizations
- To identify specific elements of library organization of particular relevance in supporting research
- To analyze strengths and weaknesses of potential organizational models.

2. METHODOLOGY

This research uses an exploratory sequential mixed methods design (**Creswell & Plano Clark, 2011**) to survey organization structures of research libraries in the specific context of library service responses to challenges presented by the contemporary e-research environment. The purpose was first to qualitatively explore dimensions of library structural designs with a small single-country sample and then to quantitatively determine the extent to which the qualitative findings generalized to a large international sample. The present paper reports the results from the first phase only, but describes the complete design to explain the rationale for the initial research.

In the first (instrument-development) phase, data were collected via institutional websites, organization charts, strategy documents, job descriptions and other public-domain sources from a purposive sample of 24 universities in the UK. The Russell Group of leading UK research universities was used as the sampling frame.1 The use of websites and web-based documents as sources of data and objects of analysis has been recognized as an appropriate method of investigation for contemporary researchers, particularly in relation to organizational research (**Bryman, 2004**). Previous studies have used this type of Internet-based research to investigate library strategy documents, either as the sole method of data collection (**Corrall, 2007**; **Pacios, 2004**), or as the first phase of a multi-method study (**Aldrich, 2007**; **Corrall, 2008**; **McNicol, 2005**).

Both deductive and inductive strategies were used in the initial phase. A review of related literature informed the systematic browsing and keyword searching of websites, but the key outputs of the research (such as the categorizations of structural arrangements) were generated from the data analysis. Evidence was captured and recorded electronically as raw data (document extracts or whole documents), notes and initial codes. The data were analyzed using matrix analysis techniques, using color coding to highlight examples with striking features, which were then re-examined to select cases illustrating sharp contrasts (**Corrall, 2007**). Data matrices are particularly suited to making large amounts of qualitative data accessible and meaningful, enabling insights to emerge, and supporting both cross-site and within-site comparisons (**Miles & Huberman, 1994**; **Nadin & Cassell, 2004**).

The findings from this initial exploration will be used to develop assessment measures that can be administered to a large sample. In the planned second phase, data on service configurations, reporting relationships, divisions and specializations, coordination and integration will be collected from research library directors in Australia, New Zealand, North America, and selected European countries, using an online questionnaire survey. The reason for collecting qualitative data initially was that there is little guiding theory on the subject to inform the development of a survey instrument (**Creswell & Plano Clark, 2011, p. 86**).

3. RESULTS

Formal organization structure charts were found via web searches in eight cases, i.e., for one-third of the sample investigated (*n*=24). However, relevant dimensions and salient features of the structures in use at other sites were identified by browsing their web pages and consulting staff directories, service descriptions and other information available (including structure charts for the institution in some cases, and further particulars for job announcements in a few instances). The main elements examined were service configurations, reporting lines, internal organization, specialist positions, and coordination mechanisms. Throughout the analysis, the focus was specifically on elements of library organization of particular relevance in supporting research.

The headline findings from the data are as follows:

- Integrated library and computing services have fallen from favour
- Reporting lines for library and information services have shifted
- Library structures continue to mix function and market/subject elements
- New senior positions and titles are signalling strategic priorities
- Subject liaison arrangements are being enhanced and remodelled

- Specialist research positions are emerging as a significant trend
- Dedicated spaces for researchers are improving service visibility
- Groups and websites are promoting boundary-spanning services.

3.1 Service configurations

One-third of the services featured some form of convergence between the library and one or more cognate professional services, most commonly the computing/information technology (IT) service, which often also includes learning/education technology. This finding contrasts with the position a decade ago when more than half of the 20 institutions then in the Russell Group were converged and one-quarter had merged structures with fully integrated library and IT provision for several core functions (e.g., help desk, training). Among the currently converged organizations, four had retained the same basic structure for a decade or more, of which one was still a merged service, two were converged, and one was federated (separate organizations within an information services division of professional services).

Two organizations had deconverged their library and IT services and combined the library with other cognate services to form an academic services directorate and a student services directorate respectively. Two other institutions had formed newly converged services in the last five years, in both cases bringing the library and IT together, into an information directorate in one case, and academic services in the other (which also included student services and quality enhancement). However, the key finding here is that most members of the group now have separate library organizations, which includes four high-profile services that have deconverged their merged or converged service organization within the last three years (King's College London, the London School of Economics (LSE), and the Universities of Bristol and Nottingham).

The variety of service configurations was a feature of UK converged information service organizations in the 1990s and 2000s, and this pattern has continued as services have de-converged and re-converged more recently. For example, at both the University of Birmingham and the University of Exeter, Library services is one division within an Academic Services directorate of seven divisions, but the composition of each directorate and structure of the library division varies. Table 1 shows the breakdown (based on data collected in summer 2013).

Table 1. Sample composition of academic services directorates

University of Birmingham		University of Exeter	
Academic Services Divisions	*Library Services Departments*	*Academic Services Divisions*	*Library & Culture Services Groups*
• Birmingham Foundation Academy	• Academic Liaison, Learning, Research Support & Academic Skills	• Academic Support Unit (Cornwall)	*Library & Research Support Teams*
• Centre for Learning & Academic Development and Learning Spaces	• Collection Management & Development	• Education Quality & Enhancement	• Academic Engagement
• Student Employability	• Library Customer Support	• Employability & Graduate Development	• E-Strategy and Resource Development
• Library Services		• Exeter IT	• Heritage Collections
• Special Collections		• Library & Culture Services	• Library Customer Services
• Registry		• Planning & Resources	*Arts & Culture Team*
• Student Services		• Student Services	

At divisional level, in addition to library services, common functions here include academic development/educational enhancement, student employability (careers services), and student services/registry functions (combined at Exeter). However, technology services (including administrative systems, academic computing and student IT services) are part of the converged service at Exeter, but at Birmingham, in the re-converged service, IT Services is a separate unit within Professional Services reporting to the Registrar & Secretary – unlike the former converged information service organization at Birmingham, where IT was a central part of the original unified structure. At the library divisional level, Table 1 shows that the primary split into four functional sections is similar in both institutions, namely: academic liaison/ engagement; customer services/support; collection/resource development; and special/heritage collections. However, Birmingham has departed from established practice by making Special Collections a separate division (not reporting to the Director of Library Services), while Exeter has brought in Arts and Culture as an additional service group reporting to the Assistant Director of Academic Services/Head of Library and Culture Services.

The make-up of the academic support department/team within each library also differs slightly. The Birmingham department includes responsibility for the University's Academic Skills Centre (which provides support in areas such as academic writing, maths and stats, and study skills), in addition to academic liaison, and research support (including supporting the REF and research data management); whereas Academic Skills Development at Exeter is located within the Education Enhancement division, and the Library's Academic Engagement group includes separate teams specializing in digital content and library marketing, and open access and data curation, in addition to liaison support for teaching and research. (At Birmingham, responsibility for managing and supporting the institutional repository and open access is part of the Collection Development and Management department.) Both institutions use the traditional subject librarian model to carry out liaison activities: at the time of data collection, Birmingham had 15 Subject Advisors supporting 5 Colleges, while Exeter had 10 Subject/Liaison Librarians supporting 4 Colleges and separate Business and Medical Schools (part of the College structure at Birmingham). The numbers of liaisons are commensurate with the respective sizes of the populations served, Birmingham being approximately 50 percent larger than Exeter.

The Information Services directorates that have continued as merged, converged or federated services are now more uniform in their service portfolios. All include libraries (and archives), IT (academic computing and corporate systems), and learning technologies, usually including responsibility for both equipping and managing teaching spaces, but not covering academic skills (other than information literacy), which are typically provided by academic support/learning development units, nor other student services functions. Within the sample, although some instances of shared support functions were found (e.g., Operations and Central Projects at Birmingham; Strategic Support, Finance, HR, etc. at Cardiff; Planning and Resources at Exeter), only Edinburgh currently displays elements of the more radical merged service structures that were adopted to varying extents by Birmingham and other leading research universities from the mid 1990s.

The key features of the Edinburgh structure are the combined library, e-learning and IT first-line help/support service, and the organization of liaison librarians and IT professionals into three College-facing consultancy teams within the User Services Division. The consultancy teams are in a different part of the organization from other library staff, who are mainly located in the Library and Collections Division. This means they are separated from staff working in related research and learning support services, including teams and sections dealing with scholarly communication, and collection development and management, and also do not report to the *de facto* library service head, the Director of Library & University Collections.

3.2 Reporting relationships

In UK universities, the Librarian traditionally reported to the Vice-Chancellor or Principal (as the chief executive officer of the institution), which often differentiated the library from the IT/computing service; but the adoption of more formal management arrangements in higher education institutions over the past two decades and the reconfiguration of administrative support functions into groups of professional services has resulted in different structures. Nearly three-quarters of the library directors in the sample report directly (or indirectly, via another senior officer) to the University's Registrar/Secretary/Chief Operating Officer. Chief Operating Officer (COO) is becoming the dominant title in use, found in half of the institutions.

Only Oxford and Cambridge seem to have retained their traditional reporting line in its original form, although the head of the converged service at Edinburgh (who holds the formal title of Vice-Principal for Knowledge Management, CIO and University Librarian) reports directly to the Principal. At the time of data collection, the university librarian/library director at Manchester and Imperial reported to the equivalent of the Deputy Vice-Chancellor (Deputy Rector at Imperial); at Leeds and Nottingham, they reported to a Pro-Vice Chancellor (PVC), in both cases the PVC with responsibility for education/teaching and learning. The grouping of the library with education/student services is another notable trend in structural arrangements, reflected in the recently converged academic service model adopted at Exeter, the reconverged academic services directorate at Birmingham, and reconverged student services at Queen Mary (University of London). In addition, the deconverged library service at King's now reports to the College Secretary via the Director of Students & Education; similarly at Bristol, the reporting line is via the Deputy Registrar, responsible for "the student-facing and education-related services".

It is debatable whether a PVC reporting line is preferable to the COO line, particularly where the PVC portfolio has a focus on teaching/students. Reporting directly to the COO role can position the library director alongside other professional service heads, including both student services and research services, which is particularly valuable in the context of collaborative service development with the research office or equivalent in areas such as managing research data and measuring research impact.

3.3 Internal organization

The size and shape of the library superstructures vary significantly, with leadership teams ranging from three to eight members. The titles of senior management positions and labels used for the main divisions/departments/operational units also vary, and some libraries combine one or more functions together, but overall there was a striking consistency in the way activities were grouped into the following five areas:

- Academic services/research and learning support (generally subdivided by subject field)
- Collection development/information resource management
- Special collections and archives/heritage collections
- Customer services/reader services/user support
- E-strategy/digital services/information technologies.

The academic (learning and research) liaison/support function was explicitly represented in the senior team in three-quarters of the sample. In one-third of these (six cases), the relevant manager had other responsibilities (e.g., Planning & Resources, Collection Development/Management, Special Collections, User Services, Service Development). In one novel example, at the University of Manchester, research services, teaching and learning support, and academic engagement were distributed among three different members of the leadership team, and combined with other related responsibilities. At Edinburgh, the merged Information Services organization separates consultancy including library liaison services (in the User Services Division) from other library-based research and learning services (in the Library & Collections Division).

Academic Services was the descriptor used most consistently, found in the job titles of seven senior managers. However Research, or Research

& Learning (or Learning & Research) featured in seven titles of associate/assistant directors/heads of division or equivalent. The word Liaison was included in four titles, and the term Academic Engagement in two jobs. Other variants included the position of Head of Relationship Management at York, and the five Heads of Faculty Services in the senior team at Southampton (where academic liaison is directed by the Deputy Librarian). The most complex title identified was Assistant Director – Academic Liaison, Learning & Research Support, and Academic Skills at the University of Birmingham (a position that has subsequently been changed to Academic Engagement).

The collection development/management function was equally prominent, also being represented in the senior teams of three-quarters of the sample. However, the term "collection" was used explicitly in slightly less than half the sample, and the titles of the relevant roles, divisions, or sections also varied within these cases, e.g., Collection & Access Management (Nottingham), Collection Development & Description (Cambridge), Collection Services (Durham, Leeds). The term "resources" seems to be gaining currency as an alternative to "collection", suggesting the concept of the library collection may become less prominent as migration from purchased printed materials to licensed e-journals and e-books continues. At York, the relevant senior position is Head of Content.

In approximately half of the cases featuring a collection development or management role in the top team, the collections function was combined with the management or strategic development of library systems (in two cases along with facilities), represented by roles such as Head of Collections and eStrategy (Sheffield – a position that has now evolved into Assistant Director for Academic Services by adding responsibility for the Learning & Research Services team to the Procurement & Content and Digital Services & Systems teams), Head of eStrategy & Resource Development (Exeter – a position that is now combined with academic liaison responsibilities in an enlarged Academic Support portfolio), and Head of Information Management &

Library Development (Bristol). A slightly less prevalent, but equally significant, trend (found in around one-quarter of the total sample), was the creation of senior roles solely concerned with the library's information systems/information technologies portfolio, notable examples here including: Associate Director, Digital Library Programmes & Information Technologies (Oxford), Head of Digital Library Services (Leeds), and Head of Digital Technologies & Services (Manchester).

Around half of the institutions had special collections and/or archives represented as a distinct role in the library's senior team at the level of divisional/departmental head, including two who used the alternative title of Head of Heritage Collections (Durham and Exeter). Also in around half of the sample, delivery of frontline services to users (such as lending/circulation and enquiry/help services) was represented explicitly in the senior group, e.g., Assistant Director – Customer Services (Kings), Head of Library Customer Services (Exeter). In several other cases, customer/user services was part of a larger portfolio for a member of the executive. Customer Services (or Customer Support) was by far the most common label used for this function, followed by four instances of User Services, two of Reader Services, and only one where the term Public Services was used.

3.4 Subject liaison

The importance of subject-based academic support was evident in the structures of all 24 libraries in the sample. With the exception of the Universities of Cambridge (where the library's strategic objectives include development of "integrated, subject-based provision for teaching and learning as well as research") and Manchester (where academic liaison through teams of faculty librarians was recently discontinued), every institution currently uses some form of subject librarian arrangement to provide specialized support to students and academic/research staff. Seven use the title *Subject* Librarian (or Subject Advisor or Subject Consultant) for this role, while 11 use *Liaison*

Librarian (or Academic Liaison Librarian, or Faculty Liaison Librarian). (Other titles found include Academic Support Librarian, College Team/Support Librarian, Faculty Team Librarian, and Information Specialist.) In ten cases, the library had a team structure mirroring the institutional structure of Colleges or Faculties, with subject/liaison librarians led by Heads of Faculty Services, Faculty Team Leaders, Faculty Librarians, or similar. Some libraries had additional structural differentiation by including *Assistant* Liaison Librarians, or Liaison *Assistants* in their teams.

The strength of liaison teams/subject support varied across the sample from six liaison librarians at Queen Mary (University of London) and eight at Durham, to 32 subject librarians at Cardiff and 52 at Oxford (at the time of data collection). Although generally the liaison librarian title was associated with smaller-scale provision, this was not universally the case, as the Queen's (Belfast) structure included only eight Subject Librarians (led by three Faculty Librarians), whereas Southampton had 24 Liaison Librarians, led by five Heads of Faculty Services (who also fulfilled liaison roles). The mean number across the sample was 15.9 and the median was 13.5. However, these figures should be seen as approximations only. They do not always represent the total number of librarians involved in subject-based learning support and research services, as many libraries have recently established specialist positions that focus solely on learning or research support, or a particular aspect of academic support. In addition, the method used to collect data on subject liaisons did not differentiate between full-time and part-time liaison staff.

3.5 Specialist positions

As reported in the Introduction, support for research has traditionally been a formal part of the subject/liaison/faculty librarian role, alongside support for teaching and learning, but published literature indicates a trend towards libraries putting more effort into the delivery and promotion of support for researchers, having identified opportunities to provide more specialized "higher-end" services, and also become aware that researchers often have a limited view of what librarians provide (tending to focus on resource procurement and collection management). A significant development in the past decade has thus been the creation of new roles within the academic services/support area designed to provide specialist support for researchers in areas such as literature searching, systematic reviews, current awareness, reference management, citation analysis, publishing advice, open access and data curation.

One-third of the sample libraries had established a specialist position with a research support focus, mostly within the past five years. In some examples, the role is similar to and positioned alongside existing subject/liaison librarians, with titles such as Academic Liaison Librarian (Researcher Support) at Durham, or Research Support Librarian at Queen Mary. In most cases, the post-holder is also expected to fulfill a management coordination or strategic leadership role, both carrying out specialist work and providing direction for research-related service developments across library teams; examples of relevant job titles here include Research Support Leader at Liverpool, Library Liaison Manager (Research) at King's, Head of Scholarly Communications at Southampton, and Academic Support Manager (Research) at Warwick – where the post-holder is supported by a dedicated Academic Support Librarian. In a few instances, the post-holders also act as subject librarians/academic liaison contacts (e.g., Senior Consultant, Research Support at Cardiff, and Research Support Services Manager at LSE).

In addition to this type of cross-cutting coordination and development role, several libraries have established more specific research support positions, whose remits have evolved in line with additional needs identified. For example, the Research Skills & Development Librarian at Cambridge directs the University Library's Research Skills Programme, but is also a member of the University's Open Access Project team; at Imperial, the intended focus of the new position of Research Support

Librarian (Medicine) was mediated literature searching for systematic reviews, but the role developed to provide support in other areas, such as data management. Other longer established examples of specialist research support provision in libraries include two largely grant-funded units, the Support Unit for Research Evidence (SURE, established in 2000) at Cardiff, which specializes in systematic reviews of the literature; and the Centre for Research Communications at Nottingham (established formally in 2009, but with a history going back to the SHERPA project of 2002–2006), which carries out research and development work in scholarly communication, open access and related areas.

Management of the institutional repository/research archive is now an established library function, but its location within the organization structure varies. In two instances within the sample, the repository function had been subsumed within the institutional research information system managed by the research office; in another two examples, the repository service was a joint venture with the university computing service, but located in the library. However, repository management was mostly a library responsibility, generally located either in the library's systems/IT/digital services division (10 examples), or with academic/research and learning services (9 examples). In two cases the research support librarian or equivalent was formally designated as the institutional repository manager, and in another two cases, the research support leader had formerly filled the role of repository manager. At Birmingham, the repository is the responsibility of the Collection Development and Management department, where the Digital Assets team is located (in the absence of a library IT department).

Policy developments and funding mandates related to research assessment, open access and data management have resulted in many new and redesigned specialist library positions and teams to manage and coordinate institutional support in these areas. For example, Edinburgh has a Scholarly Communications Team of four people providing support for policy development and implementation, open access advice and advocacy, intellectual property rights, bibliometrics, repository administration, publications submission, and metadata quality. Exeter has an Open Access and Data Curation Team, comprising a Manager, a Data Curation Officer, and an Open Access Administrator. At Bristol, an institutional Research Data Service was launched in summer 2013, as a Library-led collaboration with IT Services and Research & Enterprise Development; it is staffed by a Manager, two Senior Research Data Librarians and an Assistant Research Data Librarian. Sheffield has formed a new Research Support Team within its Learning & Research Services group, which includes two Coordinators for Open Access and Research Data Management, two Open Access Assistants and a Research Data Management Liaison Assistant.

3.6 Dedicated spaces

The University of Warwick Library launched its Wolfson Research Exchange in 2008 as an innovative suite of specialist services and dedicated technology-rich spaces for researchers, with provision for quiet study, interdisciplinary networking, academic meetings, and social activities, and a programme of events on topics such as funding opportunities, grant applications, data management, journal impact, and spinout companies, "offered not only by the library but by other university departments supporting research too" (**Carroll, 2011b, p. 55**). Novel services include Research Match ("a unique service which profiles your research and makes it visible") and Collaborative Sandpits (interactive multi-day workshops to create project proposals for funding).

With separate programming for graduate research students, early career researchers, and academics/research staff, the facility has its own dedicated staffing, with a Research Exchange and PG Hub Manager, a Coordinator, Support Officers, and Assistants, who work alongside other managers and academic support specialists in the Academic

Services division of the Library. *In situ*support is provided by a team of part-time Advisers, who are all current or recent PhD students (**Carroll, 2011a**). The Library website presents the facility as "A space, a website, a community", reflecting its several dimensions, including its hosting of networking evenings, interdisciplinary special interest groups, online peer to peer discussions, and blogs.

The development of dedicated purpose-designed space for researchers emerges as a significant trend creating physical presence for renewed efforts by libraries to position themselves as research partners and collaborators. The new research spaces typically have card-controlled entry to restrict access and demonstrate that libraries understand that researchers want and need physical facilities that are separate from undergraduates. Queen Mary opened a new 58-seater Research Reading Room for graduate researchers and academic staff in 2011, after a consultation identified the need for a quiet, comfortable working environment away from busier noisier parts of the Library. The facility was "designed along similar lines to The British Library Reading Rooms, with high specification study desks finished in oak and leather, providing larger than standard worktop space, and LED task lighting at each desk". Birmingham, similarly, has a Researchers' Room "reserved for use by postgraduate research students and academics only".

Social space and technology provision generally differentiate contemporary researcher facilities from traditional reading rooms. For example, Durham is developing a researchers' study area for the exclusive use of postgraduate research students, with individual, collaborative and social areas, which is also being used to promote information about library support for research. At Exeter, the Old Library housing the University's special collections has been refurbished and renamed the Research Commons, with a new large reading room, enhanced IT, multi-media seminar room, and café-style break out space and terrace area. At York, where the Library was able to extend into an adjoining building, renovation has created new improved research space, including a study area, specifically for researchers; an informal research lounge area; and – the most unusual marketing feature – a Research Hotel, a suite of bookable rooms that visiting academic staff or researchers can reserve for their exclusive use as temporary work areas. The website explains that "The hotel concept does not extend to the provision of sleeping arrangements, but does offer use of the research lounge".

3.7 Boundary-Spanning Activities

Policy developments in the UK, notably the Engineering and Physical Sciences Research Council Policy Framework on Research Data (**EPSRC, 2011**) and the Research Councils UK (**RCUK, 2013**) Policy on Open Access have given rise to other structural arrangements, such as steering/working groups bringing institutional stakeholders together to facilitate compliance with funder requirements. Such groups are typically led or chaired by a Pro-Vice Chancellor for Research or senior research administrator and usually include representatives of academic schools/faculties/colleges in addition to key staff from the library, IT service and research office or equivalent. For example, at Cambridge, an Open Access Working Group was set up in 2012 to draft an institutional policy and make recommendations on implementation, followed by an Open Access Project Board in 2013 "to manage the transition to Open Access". At Exeter, "an Open Access and Research Data Management Policy Task and Finish Group was set up in March 2012 to develop an institutional policy on research data and Open Access issues". Leeds, University College London (UCL) and Warwick all have a two-tier structure for managing research data, with a high-level steering/oversight group or project board complemented by a Research Data Working Group at Leeds, a Technical & Operations Group at UCL, and an Action/Implementation Group at Warwick; the UCL structure also includes a User Group, which meets less frequently.

Institutional websites offer another opportunity to coordinate, focus and promote services to support research and researchers. Three-quarters of the library and information services in the sample had a set of web pages bringing together information targeted at researchers. At one-third of the sites, there was a prominent link from the library/information services home page, labeled Research Support, Researchers, or simply Research; at another eight, users could reach the relevant page with two clicks. The web pages varied in the topics featured, but there was a clear trend towards selectivity and integration, bringing together resources judged most useful to researchers, offered both by the library and other institutional units (such as research computing/IT services and the research office), and also trying to present the information from a research user viewpoint, rather than a service provider perspective. For example, at King's College London, the Library Research Support home page has the sub-title "Support through the research lifecycle" and a choice of six categories to click: Ideas, Research funding, Managing information, Disseminating research, Evaluating research, and Preserving research; there is also a link to the Library Research Support site from the institutional Research Support website. The Researcher@Library website at Leeds "gives researchers at the University of Leeds a single point of access to information, support and resources across the University".

Research impact, open access, and data management feature strongly on these web pages, reflecting current preoccupations of the UK research community, and also demonstrating how libraries are taking advantage of recent policy developments to position themselves as central to the research process. In almost all institutions in the sample, the library has been assigned responsibility for managing the funds allocated for Article Processing Charges incurred in compliance with funder requirements for gold open access. Most libraries have developed new web pages to support this process, typically including tables and/or decision trees/flow charts adapted to show how the institution is interpreting and implementing the relevant policies. Web pages supporting research data management were found at two-thirds of the institutions, with half of the sites clearly branded as part of the library/information services offer. Five services (including two projects led by the library) were presented as university-level offerings, usually acknowledging contributions by different services, and two services were based in the IT service. The websites were at various stages of development, with two consisting solely of links to external sites.

4. DISCUSSION AND CONCLUSIONS

4.1 Convergence

The move away from the converged service model is one of the striking findings of the study. However, the distinction between "organizational" and "operational" convergence (**Field, 2001**, p. 268) is important here. *Organizational*convergence (i.e., formally combining two or more services to form one organization) is evidently in decline within the sample, especially structural designs merging libraries with computing/ IT services. *Operational* convergence (i.e., separate services/ departments collaborating to coordinate their activities to improve conference and effectiveness), in contrast, is arguably more prevalent than ever, with libraries extending and deepening their collaborations and partnerships beyond IT and educational development colleagues to other professional services, such as research offices. This is particularly evident in:

- services (e.g., institutional repositories, research data services) delivered as collaborative ventures with IT services and/or research offices
- relationships and structures (e.g., committees and groups) supporting boundary-spanning activities in areas such as data management and open access
- unified research support websites providing a single point of access to university-wide resources for researchers, which could be

described as an example of a *virtually* converged information service.

4.2 Positioning

The location of the library/information service organization in the institutional organizational structure is rarely discussed in the literature, but is important because it has implications for the allocation of resources, the coordination of activities, and the perception of the service by stakeholders. The two significant trends here were first, the administrative positioning of the library within professional services reporting to the Chief Operating Officer or equivalent, which was identified as the dominant model; and secondly, an increasing tendency to group libraries with "student-facing and education-related services" through reporting lines and directorate structures. The latter is arguably not a positive move for libraries trying to reposition themselves as credible collaborators and partners in research activities.

4.3 Specialization

The survey showed that departmentalization within libraries has evolved beyond the traditional reader/technical services model to a focus on five strategic areas of activity: information resources, academic engagement, customer service, heritage collections, and digital technologies, although there was significant variation in the size of leadership teams. Use of the term "Research" in the job titles of around one-third of the senior management positions identified sends a signal to stakeholders about the priority attached to this area. The strong focus on information systems/technologies identified at the senior management level is not surprising, but it was interesting to find that in many cases the technology portfolio was combined with another area of responsibility, and most frequently with information resources/

collections, which in effect continued the tradition of locating systems librarians in a technical services division/department.

Specialization of library professional staff continues to focus on the subjects/academic disciplines taught and researched in the institution, but subject/liaison roles are increasingly being complemented and supplemented by coordinating roles (such as research support managers) and functional specialists for particular areas of activity (e.g., research data management, open access publishing). The hybrid approach to specialization found in UK research libraries echoes structural developments reported in North America (**Covert-Vail & Collard, 2012**; **Jaguszewski & Williams, 2013**).

An interesting dimension of specialization is the growing trend to extend the development of specialized research support into dedicated spaces for researchers. The space-as-service concept is gaining currency internationally as a visible contribution to research, particularly in support of early career researchers, and interdisciplinary collaboration, with the research commons beginning to emerge as a parallel development to the successful learning commons movement (**Corrall & Lester, 2013**; **Daniels, Darch, & de Jager, 2010**;**Ohio State University Libraries, 2012**). UK examples range from traditional reading rooms to the fully-serviced technology-rich space at Warwick, using early career researchers as peer support advisors, and offering novel services such as research profile matching and suites of bookable rooms for individuals and meetings (**Carroll, 2011a,b**).

4.4 Coordination

Coordination within libraries is being formalized through roles such as research support managers and leaders, and data management and open access coordinators (and also by internal committees and working groups). Coordination with other services and operations is being pursued through and facilitated by institution-wide groups at both strategic and operational level, typically established to deal with external mandates, but offering the potential of additional benefits for librarians in relationship building, professional credibility, and institutional positioning. As noted above, websites are an additional valuable way of coordinating research services to provide a single point of access for researchers and can be used not only to pull together library and other institutional resources and services, but again to position the library as a major player in the research arena. Research support websites complement the work of research support coordinators in libraries, but can also fulfill that role in libraries which do not have an individual tasked with that responsibility.

4.5 Outcomes

The purpose of the present study was to explore qualitatively how a sample of academic research libraries are organizing resources and services to support their research mission with the intention of identifying important dimensions of organizational design that could be used to develop an assessment instrument for a larger scale international survey. The study achieved its objectives and the results both confirm and extend the findings of related research investigating library plans, roles and skills for supporting and serving researchers. Although the sample was limited to 24 libraries in one country, the results identify structural variables related to library support for research that can be used as a provisional framework for libraries in the UK and other countries to examine and evaluate their own structures.

The present study was conceived as the first phase of a larger scale international survey of organizational design in research libraries and the findings are being used to develop a questionnaire to collect quantitative data on important structural and contextual variables identified from this exploratory study and ongoing review of related literature. The dimensions to be investigated include:

- positioning of libraries within their parent institutions
- scope of library/information service organizations
- size and focus of library leadership teams
- specialization of units, roles and space
- formal and informal convergence of activities
- mechanisms facilitating boundary-spanning activities.

NOTES

[1]http://www.russellgroup.ac.uk.

ACKNOWLEDGMENTS

A summary of this study was presented at the 2013 LIBER Annual Conference in Munich, Germany. A synopsis of the key findings was also published in *Library Connect Newsletter, 11*(3), The Library and the Research Enterprise (pp. 3–4).

REFERENCES

1. Adema, J., & Schmidt, B. (2010). From service providers to content producers: New opportunities for libraries in collaborative open access book publishing. New Review of Academic Librarianship, 16(sup. 1), 28–43.
2. Aldrich, A.W. (2007). Following the phosphorous trail of research library mission statements into present and future harbors. ACRL

13th National Conference, March 29–April 1, 2007, Baltimore, MD (pp. 304–316). Chicago, IL: Association of College & Research Libraries. Retrieved April 30, 2014, from http://www.ala.org/acrl/sites/ala.org.acrl/files/content/conferences/confsandpreconfs/national/baltimore/papers/304.pdf

3. Andrade, R., & Zaghloul, R. (2010). Restructuring liaison librarian teams at the University of Arizona Libraries, 2007–2009. New Library World, 111(7/8), 273–286.

4. Auckland, M. (2012). Re-skilling for research: An investigation into the role and skills of subject and liaison librarians required to effectively support the evolving information needs of researchers. London: RLUK Research Libraries UK. Retrieved April 30, 2014, from http://www.rluk.ac.uk/wp-content/uploads/2014/02/RLUK-Re-skilling.pdf

5. Bourg, C., Coleman, R., & Erway, R. (2009). Support for the research process: An academic library manifesto. Dublin, OH: OCLC Research. Retrieved April 30, 2014, from http://www.oclc.org/content/dam/research/publications/library/2009/2009-07.pdf

6. Bradbury, K., & Weightman, A. (2010). Research support at Cardiff University Library. SCONUL Focus, 50, 65–70. Retrieved April 30, 2014, from http://www.sconul.ac.uk/sites/default/files/documents/19_2.pdf

7. Brunner, M.L. (2010). PhD holders in the academic library: The CLIR Postdoctoral Fellowship Program. In K. Williams & S. Walter (Eds.), The expert library: Staffing, sustaining, and advancing the academic library in the 21st century (pp. 158–189). Chicago, IL: Association of College & Research Libraries. Retrieved June 23, 2014, from http://escholarship.org/uc/item/05j228r4

8. Bryman, A. (2004). Social research methods. (2nd. ed.). Oxford: Oxford University Press

9. Bryson, T., Posner, M., St. Pierre, A., & Varner, S. (2011). Digital humanities: SPEC Kit 326. Washington, DC: Association of Research Libraries

10. Bulpitt, G. (Ed.) (2012). Leading the student experience: Super-convergence of organisation, structure and business processes. London: Leadership Foundation for Higher Education. Retrieved April 30, 2014, from http://www.lfhe.ac.uk/en/components/publication.cfm/S3%20-%2005

11. Carlson, J., & Kneale, R. (2011). Embedded librarianship in the research context: Navigating new waters. College & Research Libraries News, 72(3), 167–170. Retrieved April 30, 2014, from http://crln.acrl.org/content/72/3/167.full

12. Carroll, D.L. (2011a). Fostering a community of scholars at the University of Warwick: The Wolfson Research Exchange. New Review of Academic Librarianship, 17(1), 78–95. doi: 10.1080/13614533.2011.547820. Retrieved June 23, 2014, from http://wrap.warwick.ac.uk/4540

13. Carroll, D. (2011b). What researchers want: A personal case study. SCONUL Focus, 51, 52–56. Retrieved April 30, 2014, from http://www.sconul.ac.uk/sites/default/files/documents/16_3.pdf

14. Chan, D.L.H., & Soong, S.C. (2011). Strategic repositioning in a dynamic environment. Library Management, 32(1/2), 22–36.

15. Child, J. (1988). Organization: A guide to problems and practice (2nd ed.). London: Paul Chapman

16. Corrall, S. (2007). Benchmarking strategic engagement with information literacy in higher education: Towards a working model. Information Research, 12(4), paper 328. Retrieved April 30, 2014, from http://www.informationr.net/ir/12-4/paper328.html

17. Corrall, S. (2008). Information literacy strategy development in higher education: An exploratory study. International Journal of Information Management, 28(1), 26–37.

18. Corrall, S., & Lester, R. (2013). The researcher's view: Context is critical. In L. Watson (Ed.), Better library and learning spaces: Projects, trends and ideas (pp. 183–192). London: Facet

19. Corrall, S., Kennan, M.A., & Afzal, W. (2013). Bibliometrics and research data management services: Emerging trends in library support for research. Library Trends, 61(3), 636–674. Retrieved April 30, 2014, from http://d-scholarship.pitt.edu/18948/.

20. Cotta-Schønberg, M. (2007). The changing role of the subject specialist. Liber Quarterly, 17(3/4). Retrieved April 30, 2014, from http://liber.library.uu.nl/index.php/lq/article/view/7890/8111

21. Covert-Vail, L., & Collard, S. (2012). New roles for new times: Research library services for graduate students. Washington, DC: Association of Research Libraries. Retrieved April 30, 2014, from http://www.arl.org/storage/ documents/ publications/nrnt-grad-roles-20dec12.pdf

22. Cox, A., Verbaan, E., & Sen, B. (2012). Upskilling liaison librarians for research data management. Ariadne, 70. Retrieved April 30, 2014, from http://www.ariadne.ac.uk/issue70/cox-et-al

23. Creamer, A.T., Morales, M.E., Kafel, D., Crespo, J., & Martin, E.R. (2012). A sample of research data curation and management courses. Journal of eScience Librarianship, 1(2), 88–96. doi: 10.7191/jeslib.2012.1016. Retrieved April 30, 2014, from http://escholarship.umassmed.edu/jeslib/vol1/iss2/4/

24. Creswell, J.W., & Plano Clark, V.L. (2011). Designing and conducting mixed methods research (2nd ed.). Thousand Oaks, CA: SAGE

25. Crow, R., Ivins, O., Mower, A., Nesdill, D, Newton, M., Speer, J., & Watkinson, C. (2012). Library publishing services: Strategies for success. Washington, DC: SPARC. Retrieved April 30, 2014, from http://wp.sparc.arl.org/lps/

26. Daniels, W., Darch, C., & de Jager, K. (2010). The research commons: a new creature in the library. Performance Measurement and Metrics, 11(2), 116–130.

27. Delserone, L.M., Kelly, J.A., & Kempf, J.L. (2010). Connecting researchers with funding opportunities: A joint effort of the Libraries and the University Research Office. Collaborative Librarianship, 2(1), 33–37. Retrieved April 30, 2014, from

http://collaborativelibrarianship.org/index.php/jocl/article/view/59/39

28. Drummond, R., & Wartho, R. (2009). RIMS: The Research Impact Measurement Service at the University of New South Wales. Australian Academic & Research Libraries, 40(2), 76–87.

29. Duncan, R. (1979). What is the right organization structure? Decision tree analysis provides the answer. Organizational Dynamics, 7(3), 59–80

30. Duranceau, E.F. (2008). The "Wealth of Networks" and institutional repositories: MIT, DSpace, and the future of the scholarly commons. Library Trends, 57(2), 244–261.

31. EPSRC (2011). EPSRC policy framework on research data. Swindon: Engineering and Physical Sciences Research Council. Retrieved April 30, 2014, from http://www.epsrc.ac.uk/about/standards/researchdata/Pages/policyframework.aspx

32. Eustis, J.D., & Kenney, D.J. (1996). Library reorganization & restructuring: A SPEC kit. Washington, DC: Association of Research Libraries

33. Fearon, D., Gunia, B., Pralle, B.E., Lake, S., & Sallans, A.L. (2013). Research data management services: SPEC Kit 334. Washington, DC: Association of Research Libraries

34. Federer, L. (2013). The librarian as research informationist: A case study. Journal of the Medical Library Association, 101(4), 298–302.

35. Field, C.D. (2001). Theory and practice: Reflections on convergence in United Kingdom universities. LIBER Quarterly, 11(3), 267–289. Retrieved April 30, 2014, from http://liber.library.uu.nl/index. php/lq/article/view/7646/7682

36. Franklin, B. (2009). Aligning library strategy and structure with the campus academic plan: A case study. Journal of Library Administration, 49(5), 495–505.

37. Fonseca, A.J., & Viator, V.P. (2009). Escaping the island of lost faculty: Collaboration as a means of visibility. Collaborative Librarianship, 1(3), 81–90.

38. Furlough, M.J. (2010). The publisher in the library. In S. Walter & K. Williams (Eds.). The expert library: Staffing, sustaining, and advancing the academic library in the 21st century (pp. 190–233). Chicago, IL: Association of College and Research Libraries. Retrieved June 23, 2014

39. García-Milian, R., Norton, H.F., Auten, B., Davis, V.I., Holmes, K.L., Johnson, M., & Tennant, M.R. (2013). Librarians as part of cross-disciplinary, multi-institutional team projects: Experiences from the VIVO collaboration. Science & Technology Libraries, 32(2), 160–175. doi: 10.1080/0194262X.2013.791183. Retrieved June 23, 2014, from http://www.ncbi.nlm.nih.gov/pmc/articles/PMC3700548/

40. Giesecke, J. (2010). Finding the right metaphor: Restructuring, realigning, and repackaging today's research libraries. Journal of Library Administration, 51(1), 54–65.

41. Gold, A. (2007). Cyberinfrastructure, data, and libraries, Part 2. Libraries and the data challenge: Roles and actions for libraries. D-Lib Magazine, 13(9/10).

42. Greyson, D., & Surette, S. (2013). "You're just one of the group when you're embedded": report from a mixed-method investigation of the research-embedded health librarian experience. Journal of the Medical Library Association, 101(4), 287–297.

43. Hanson, T. (Ed.) (2005). Managing academic support services in university: The convergence experience. London: Facet

44. Harris-Pierce, R.L., & Liu, Y.Q. (2012). Is data curation education at library and information science schools in North America adequate? New Library World, 113(11/12), 598–613

45. Healy, A.M. (2010). Increasing the visibility of the library within the academic research enterprise. Issues in Science and Technology Librarianship, 63.

46. Herther, N.K. (2009). Research evaluation and citation analysis: key issues and implications. The Electronic Library, 27(3), 361–375.

47. Housewright, R., Schonfeld, R.C., & Wulfson, K. (2013a). Ithaka S+R US faculty survey 2012. New York: Ithaka S+R. Retrieved April 30, 2014, from http://www.sr.ithaka.org/research-publications/us-faculty- survey-2012

48. Housewright, R., Schonfeld, R.C., & Wulfson, K. (2013b). Ithaka S+R/JISC/RLUK US survey of academics 2012. New York: Ithaka S+R. Retrieved April 30, 2014, from http://sr.ithaka.org/research-publications/ ithaka-sr-jisc-rluk-uk-survey-academics-2012

49. Jahnke, L.M., & Asher, A. (2012). The problem of data: Data management and curation practices among university researchers. In L. Jahnke, A. Asher & S.D.C. Keralis, The problem of data (pp. 3–31). Washington, DC: Council on Library and Information Resources. Retrieved April 30, 2014, from http://www.clir.org/pubs/ reports/ pub154

50. Jaguszewski, J.M., & Williams, K. (2013). New roles for new times: Transforming liaison roles in research libraries. Washington, DC: Association of Research Libraries. Retrieved April 30, 2014, from http://www.arl.org/storage/documents/publications/NRNT-Liaison-Roles-final.pdf

51. Larsen, S. & Riis, C. (2012). Aarhus University Library: Restructuring a university's library organisation. Bibliothek Forschung und Praxis, 36(2), 235–240.

52. Lee, C.A., & Tibbo, H. (2011). Where's the archivist in digital curation: Exploring the possibilities through a matrix of knowledge and skills. Archivaria, 72, 123–168.

53. Long, M.P., & Schonfeld, R.C. (2014). Ithaka S+R Library Survey 2013. New York: Ithaka.

54. Lynch, C. (2008). The institutional challenges of cyberinfrastructure and e-research. EDUCAUSE Review, 43(6), 74–88. Retrieved April 30, 2014

55. Lyon, L. (2012). The informatics transform: Re-engineering libraries for the data decade. International Journal of Digital Curation, 7(1), 126–138.

56. MacColl, J., & Jubb, M. (2011). Supporting research: Environments, administration and libraries. Dublin, OH: OCLC Research. Retrieved April 30, 2014, from http://www.oclc.org/resources/research/publications/library/2011/2011-10.pdf

57. Manchester University Library (2012). News from member libraries. SCONUL Focus,56, 61–62. Retrieved April 30, 2014, from http://www. sconul.ac.uk/sites/default/files/documents/20_6.pdf

58. McBain, I., Culshaw, H., & Walkley Hall, L. (2013). Establishing a culture of research practice in an academic library: an Australian case study. Library Management, 34(6/7), 448–461.

59. McGuigan, G.S. (2012). Addressing change in academic libraries: A review of classical organizational theory and implications for academic libraries. Library Philosophy and Practice, Paper 755. Retrieved April 30, 2014, from http://digitalcommons. unl.edu/libphilprac/755/

60. McNicol, S. (2005). The challenges of strategic planning in academic libraries. New Library World, 106(11/12), 496–509.

61. Miles, M.B. & Huberman, A.M. (1994). Qualitative data analysis: An expanded sourcebook. (2nd. ed.). Thousand Oaks, CA: SAGE

62. Miles, R.E., Snow, C.C., Fjeldstad, O.D., Miles, G., & Lettl, C. (2010). Designing organizations to meet 21st-century opportunities and challenges. Organizational Dynamics, 39(2), 93–103.

63. Nadin, S. & Cassell, C. (2004). Using data matrices. In C. Cassell & G. Symon (Eds.), Essential guide to qualitative methods in organizational research. (pp. 271–287). London: SAGE. Retrieved June 23, 2014

64. Nesdill, D., Love, A., & Hunt, M. (2010). From subject selectors to college and interdisciplinary teams. Science & Technology Libraries, 29(4), 307–314.

65. Nowviskie, B. (2013). Skunks in the library: A path to production for scholarly R&D. Journal of Library Administration, 53(1), 53–66.

66. Nutefall, J.E., & Chadwell, F.A. (2012). Preparing for the 21st century: Academic library realignment. New Library World, 113(3/4), 162–173.

67. Ohio State University Libraries (2012). Overview of research commons in academic libraries: A white paper. Columbus, OH: Ohio State University Libraries. Retrieved April 30, 2014

68. Pacios, A.R. (2004). Strategic plans and long-range plans: Is there a difference? Library Management, 25(6/7), 259–269.

69. Posner, M. (2013). No half measures: Overcoming common challenges to doing digital humanities in the library. Journal of Library Administration, 53(1), 43–52.

70. Pugh, L. (1997a). Convergence in academic support services. London: British Library Research and Innovation Centre

71. Pugh, L. (1997b). Some theoretical bases of convergence. New Review of Academic Librarianship, 3(1), 49–66.

72. Pugh, L. (2005). The management of hybrid libraries. Library and Information Research, 29(92), 13–31. Retrieved April 30, 2014

73. Pugh, L. (2007). Change management in information services (2nd ed.). Abingdon: Ashgate

74. Quinlan, C., & McHarg, H. (2012). The emerging library: Structure, culture, and lessons learned from the dissolution of a combined libraries–IT organization. Journal of Library Administration, 52(2), 147–161.

75. Radom, R., Feltner-Reichert, M., & Stringer-Stanback, K. (2012). Organization of scholarly communication services: SPEC Kit 332. Washington, DC: Association of Research Libraries

76. RCUK (2013). RCUK policy on open access and supporting guidance. Swindon: Research Councils UK. Retrieved April 30, 2014

77. Renfro, P., & Neal, J.G. (2012). The integration of libraries and academic computing at Columbia: New opportunities for internal and external collaboration. Journal of Library Administration, 52(2), 162–171.

78. Schrader, A., Shiri, A., & Williamson, V. (2012). Assessment of the research learning needs of the University of Saskatchewan librarians: A case study. College & Research Libraries, 73(2), 147–163. Retrieved April 30, 2014, from http://crl.acrl.org/content/73/2/147.full.pdf

79. Shore, E. (2012). Embracing hybridity: The merged organization, Alt/ac and higher education. Journal of Library Administration, 52(2), 189–202.

80. Stanford, N. (2013). Organization design: Engaging with change (2nd ed.). Abingdon: Routledge

81. Stanton, J.M., Kim, Y., Oakleaf, M., Lankes, R.D., Gandel, P., Cogburn, D., & Liddy, E.D. (2011). Education for eScience professionals: Job analysis, curriculum guidance, and program considerations. Journal of Education for Library and Information Science, 52(2), 79–94

82. Vandegrift, M., & Varner, S. (2013). Evolving in common: Creating mutually supportive relationships between libraries and the digital humanities. Journal of Library Administration, 53(1), 67–78.

83. Vaughan, K.T.L., Hayes, B.E., Lerner, R.C., McElfresh, K.R., Pavlech, L., Romito, D., ..., Morris, E.N. (2013). Development of the research lifecycle model for library services. Journal of the Medical Library Association, 101(4), 310–314.

84. Vinopal, J., & McCormick, M. (2013). Supporting digital scholarship in research libraries: Scalability and sustainability. Journal of Library Administration, 53(1), 27–42. Retrieved June 23, 2014, from https://archive.nyu.edu/bitstream/2451/31698/2/Vinopal_McCormick_Final.pdf

85. Warren, V. (2011). Using workforce structures to enable staff retention and development: An academic library case study. New Library World, 112(1–2), 8–18.

86. Yakel, E., Conway, P., Hedstrom, M., & Wallace, D. (2011). Digital curation for digital natives. Journal of Education for Library and Information Science, 52(1), 23–31. Retrieved June 23, 2014

87. Young, H., & Lund, P. (2008). Reflections on a benchmarking survey of research support provided by 1994 Group libraries. SCONUL Focus, 43, 51–56. Retrieved April 30, 2014

NOTES

[1]http://www.russellgroup.ac.uk.

Chapter 2

Transformational Leadership and Stakeholder Management in Library Change

Dolores Sucozhañay[1], Lorena Siguenza-Guzman[2], Cristian Zhimnay[3], Dirk Cattrysse[4], Guido Wyseure, Karel[5] De Witte[6], Martin Euwema[6]

[1]Faculty of Economics, University of Cuenca, Ecuador, and Research Groupon Occupational & Organizational Psychology and Professional Learning, KU Leuven, Belgium
[2]Department of Computer Science, University of Cuenca, Ecuador, and Centre for Industrial Management Traffic & Infrastructure, KU Leuven, Belgium
[3]ACORDES Centre, Faculty of Economics, University of Cuenca, Ecuador
[4]Centre for Industrial Management Traffic & Infrastructure, KU Leuven, Belgium
[5]Division of Soil and Water Management, KU Leuven, Belgium
[6]Research Group on Occupational & Organizational Psychology and Professional Learning, KU Leuven, Belgium

ABSTRACT

This article aims to analyse the role of library managers as change agent when implementing Library 2.0, using transformational leadership and stakeholder management approaches. To do so, a case study in a Latin

American academic library was performed. The experiences acquired for a period of six years were analysed, during which three library managers were involved in managing change. Qualitative data from documents, interviews, and observations were collected, and qualitative analysis methods were used to obtain in-depth understanding of the change process. Results show that lack of transformational leadership and stakeholder management contribute to delayed implementation and limited adoption of innovations. Although library managers recognized the importance of different stakeholders to implement changes, they did not apply systematic and proactive strategies to define and manage them. All in all, library managers should be trained as change agents, with emphasis on transformational leadership and stakeholder management skills.

KEYWORDS

academic libraries; Library 2.0; transformational leadership; stakeholder management

1. INTRODUCTION

The emergence of Internet and Web technologies is not only changing the way people search for information, but also the way information is being published and disseminated (**Jantz, 2012**). Accordingly, to maintain their status as major information providers, libraries are using new technologies for delivering their services (**vanDuinkerken & Mosley, 2011**). Library 2.0 is a revolutionary concept that focuses on user-centred changes, through the use of Web 2.0 tools such as: selective dissemination of information (SDI), forums, social networks, electronic newsletters, enhanced online public access catalogues (e-OPAC), and virtual references. However, this concept is not just about implementing technology; it implies a significant paradigm shift, requiring rethinking and changing traditional library processes and services. Unfortunately, traditional management practices, poor planning activities, lack of management focus, and lack of involvement of important stakeholders may hinder Library 2.0 deployment (**Munatsi, 2010**). Therefore, to carry out such deep changes, good leadership is key.

Riggs (2001) posits that traditionally, libraries have been over-managed and under-led. This means that although libraries have been well managed, leadership has been pushed aside and overlooked. The words "leadership" and "management" are often erroneously used interchangeably (**Riggs, 2001**). In fact, management and leadership are fundamentally different skills; management is related to performing existing practices in an adequate and efficient way, while leadership is defined as setting, motivating and aligning people with a new direction (**Kotter, 1996**). Therefore, Library 2.0, a philosophy of continuous changes, demands that leadership no longer be ignored and requires library managers or administrators to rethink the way they lead (**Riggs, 2001**). In this article, the term "library manager" is used to refer to a director or the formal position of authority.

The study of library leadership has increased significantly in the last years. Although different qualities of effective leaders have been identified, transformational leadership has been found as more effective in managing libraries coping with complex changes (**Castiglione, 2006**; **Jantz, 2012**;**McGuigan, 2012**). Transformational leadership motivates followers to go beyond their own interests and move closer towards the organizational interests (**Yukl, 2013**).

Library managers have considerable impact on library change; however, different stakeholders, such as university authorities and library staff, also have a strong influence on these changes (**López & Vargas, 2012**; **Solís & López, 2000**;**Stavridis & Tsimpoglou, 2012**). In fact, the internal culture, structure and decision-making of the academic libraries are also influenced by the 'mother' organizational culture (**vanDuinkerken & Mosley, 2011**). Nevertheless, most library leadership literature focuses on the leader-follower relationship (**Tam & Robertson, 2002**; **Williamson, 2008**), which limits the understanding of the relationships among library managers and other university actors.

The aim of this article, therefore, is to highlight the role of library managers in implementing Library 2.0, by focusing on their leadership style, and stakeholder management practices. The combination of these two essential tasks has not, to our knowledge, been studied before in this context. To do so, the experiences acquired during six years of attempted change in an academic library in Latin America have been analysed. This academic library was selected as it belongs to one of the most prestigious universities, and is considered one of the most modern and largest libraries of its country. The subsequent section recalls briefly the main theoretical ideas concerning leadership and stakeholder management in change (§2), followed by a description of the procedures used in this study (§3), leading to the delineation of its findings (§4), discussion (§5) and conclusions (§6).

2. LITERATURE REVIEW

There is a plethora of transformational leadership and stakeholder management literature from business arenas. However, according to **vanDuinkerken and Mosley (2011)**, it is still difficult to understand how to apply these insights to the library environment and, more precisely, to academic libraries.

2.1. Transformational leadership in academic library change

Bass (1990) suggests that leaders who exhibit transformational behaviours can influence followers to go beyond their own interests and move towards the organizational interests. In fact, during changes in academic libraries, transformational leadership has been found in several case studies to be more effective than other leadership styles, such as transactional and laissez-faire leadership (**Castiglione, 2006**; **Jantz, 2012**; **McGuigan, 2012**).

Transformational leaders achieve extraordinary levels of follower trust, admiration, motivation, commitment, loyalty and performance (**Kouzes & Posner, 2012**; **Yukl, 2013**), by means of four dimensions:

1. Idealized influence
2. Inspirational motivation
3. Intellectual stimulation and
4. Individualized consideration

Bass and Riggio (2006), and **Kouzes and Posner (2012)** describe these four dimensions more in detail. For instance, idealized influence is defined as serving as role models for perseverance and self-sacrifice; leaders are admired, respected and trusted. Inspirational motivation provides an appealing vision, and stipulates how the team can achieve the desired future. Intellectual stimulation encourages followers to be innovative and creative by analysing issues from a new perspective. Finally, individualized consideration means to pay attention to each individual's needs and desires, including coaching, supporting and a two-way exchange communication. **Jantz (2012)** in his study concludes that transformational styles empower librarians to create a more innovative environment, and **Castiglione (2006)** points out that organizational learning and adaptation are encouraged by transformational behaviours.

Transformational leadership is often explained in relation to transactional leadership. The latter, also known as managerial leadership is mainly defined as an exchange process in which leaders "promise and deliver rewards to employees for carrying out assignments" (**Bass, 1990**, p. 30). According to**Crosby (1996)**, transactional leadership focuses on managing the status quo with emphasis on managing subordinates by standard work, establishing well-defined processes, and managing through accomplishable tasks. Transactional leadership is characterized by two leadership behaviours: active management by exception, and contingent reward. Active management by exception is defined in terms of looking for and dealing with irregularities, mistakes, exceptions and deviations to ensure that standards are met (**Antonakis, Avolio, & Sivasubramaniam, 2003**; **Bass, 1990**). Contingent reward refers to leader behaviours that focus on both the clarification of expectations and task requirements, and offers the recognition of followers when goals are achieved (**Antonakis et al., 2003**; **Bass, 1990**).

Avolio and Bass (1991) added to their initial theory the laissez-faire or passive leadership style, which is associated with laissez-faire and passive management by exception. Laissez-faire – or "no leadership" is the absence of effective leadership, avoiding decision-making, abdicating responsibilities, and not using their authority. This leadership behaviour can be considered active to the extent that leaders choose to avoid taking actions (**Antonakis et al., 2003**). Passive management by exception is characterized by responding passively and reactively to situations and problems systematically. For instance, library managers displaying passive leadership behaviours, expect librarians to solve problems on their own without his/her direct supervision. Moreover, laissez-faire leaders do not provide a vision to be achieved by followers (**Avolio & Bass, 1991**).

Bass (1990) considers that every leader displays each of the three aforementioned styles to some extent, representing the elements of the so-called "Full Range of Leadership Model (FRL)". In this FRL model, leadership effectiveness is ranged from highly active and effective, to highly inactive and ineffective. Transformational leadership is considered highly active and effective with an emphasis on development and change (**Yukl, 2013**). Transactional leadership is situated in the middle; its effectiveness is considered in terms of focusing on irregularities, mistakes, exceptions, and deviations from standards (**Verschueren, 2014**). Laissez-faire or passive leadership style is associated with the most passive and most ineffective leadership. Both styles, transformational and transactional, can be complementary, though change effectiveness is related to greater use of transformational behaviours (**Huang & Liao, 2011**; **Yukl, 2013**).

Yukl (1999) posits that the current emphasis on the relationship between transformational leader and subordinate should be complemented with a more organizational perspective. Implementation of change implies not only influencing subordinates, but also other stakeholders at different levels (**Armenakis & Bedeian, 1999**; **Cummings & Worley, 2005**; **Govender, Moodley, & Brijball Parumasur, 2005**; **Kotter & Cohen, 2002**; **vanDuinkerken & Mosley, 2011**; **Yukl, 2013**). Thus, as **Kotter and Cohen (2002)** suggest, leaders should ensure that all actors with a stake in the change be identified and involved in the process and implementation of that change. Unfortunately, this seems not to be the case in most academic libraries, as leaders have scarce communication with stakeholders during change processes (**Stephens & Russell, 2004;vanDuinkerken & Mosley, 2011**).

2.2. Stakeholder management in academic library change

The stakeholder management approach offers an interesting frame to understand the need and the way to actively manage stakeholders in change settings. **Freeman (1984)** defines a "stakeholder" as any group or individual who can affect, or is affected by, the achievement of the change's objectives. Stakeholders are only defined in reference to a particular issue (**Bryson, 2004;Savage, Nix, Whitehead, & Blair, 1991**). In general, the stakeholder management approach involves four steps: (1) identification and analysis of stakeholders, (2) formulation of strategies, (3) facilitation of the process to implement these strategies, and (4) evaluation of strategy implementation (**Freeman, 1984**;**Mitchell, Agle, & Wood, 1997**; **Savage et al., 1991**).

Savage et al. (1991) developed a model to apply stakeholder management based on an analysis of the potential for stakeholders to threaten the process, as well as their potential to cooperate. The authors recommend that leaders should assess these two potentials and manage the stakeholders accordingly. They also suggest minimally satisfying the needs of stakeholders with low potential on both dimensions, while maximally satisfying the needs of stakeholders with high potential to cooperate, and high potential to threaten. The fundamental idea is to transform the stakeholder relationship to a more favourable condition (**Bryson, 2004**).

Library managers, as part of their role as change agents, should implement this approach. However, leaders often only identify and describe stakeholders but fail to successfully manage them (**Freeman & McVea, 2001**). Stakeholder management should be seen as a continuous task of analysing and integrating multiple stakeholders and multiple objectives (**Freeman & McVea, 2001**). Furthermore, the criteria used to assess stakeholders are not static but rather dynamic. Levels of power, saliency, coalitions and interest, may change over time, and leaders must pay attention to these changes (**Freeman, 1984**).

Concisely, whereas transformational leadership is directed at the library staff, stakeholder management focuses on other relevant parties. Based on the characteristics of transformational leadership mentioned before, it is likely that library managers with more transformational styles will also deploy more stakeholder management.

3. METHODOLOGY

A single case study concerning a six-year attempt to implement Library 2.0 into a Latin American academic library was used. This allows for a better understanding and interpretation of the events, leadership behaviours, and relationships between the library manager and the different stakeholders during the change process.

Data collection was accomplished through direct observation, analysis of secondary documents, and in-depth semi-structured interviews to assure data triangulation (**Yin, 2009**). This data collection was conducted by the first author over the six year period in order to appropriately cover the whole process and to avoid errors caused by a prospective memory failure. Through sustained observation, deep level understanding of the process was achieved. Forty official and unofficial documents were collected, including change plans, meeting reports, communications, change process reports, library regulations, strategic plans, university regulations, and university newsletters. Comprising the data were fifteen semi-structured interviews with key informants: library managers (3 interviews), librarians (9), library council members (2) and university's top administrators (1). The first round of interviews started in the second year, and continued yearly thereafter. Each interview lasted approximately 35–60 minutes. The interviews were recorded and then transcribed for analysis. The topics explored in the questionnaire, as listed in Appendix 1, were the change process, change´s results, leadership behaviours, leader relationships, stakeholder management, and limitations for change implementations.

The data analysis was carried out in several steps, mainly using qualitative software (ATLAS.ti). Firstly, data were studied and categorized meticulously in order to identify and describe the change periods and results. Codes and subcodes were used to organize the data in each period. For instance, the code "CP_Time line" was used to code the milestones of the process, and the codes "CP_First Period", "CP_Second Period", "CP_Third Period" and "CP_Fourth Period" were used to organize the events in each period. In addition, the codes "CP_Planning", "CP_Objectives", "CP_Implementation", "CP_Results" and "CP_Evaluation" were utilized to code and analyze the steps followed to implement the change, as well as the outcomes achieved in each phase. (See details in Appendix 2.) Secondly, a priority list of codes, based on our previous literature review (§ 2) and emergent ideas from the data, was used for coding. A detailed list of codes is given in Appendix 2. To ensure accuracy of coding, data were read and independently coded by two of the authors, and then were jointly compared in order to consolidate the codification. Quotations, which had been coded differently, were discussed in detail until consensus was achieved. Thirdly, matrices were developed in order to summarize and clarify codes and levels of abstraction. For more information about the use of matrices see **Miles and Huberman (1994)** and **Swanborn (2010)**. An example of a summarized matrix is given in Appendix 3. Finally, exploring data tools of Atlas.ti and matrices were used to generate findings. These emerging findings were discussed by the authors, and continuously compared and contrasted with literature on transformational leadership and stakeholder management in order to identify its emergent theoretical contribution to organizational change (**Jorgensen, 1989**).

Research quality checks were accomplished using three different procedures. First, independent coding was used. Second, the interview findings were triangulated with findings from the secondary documents and observations, avoiding dependence on the interviewees' subjective views. Third, this study employed a member checking process.

4. THE ACADEMIC LIBRARY CHANGE PROCESS

The main objective of the change process was to achieve the transition from a traditional library to a Library 2.0. The key drivers for starting this change were both the pressures of the changing context, and the implementation of an Institutional Change Project (ICP) at the university level with financial support from an external institution. The ICP´s leader was the Vice Rector, responsible for all academic affairs including academic programs, library services and students' well-being.

The six years of change were divided in four key periods related to important milestones and shifts in library manager (see Figure 1). Subsequently, the findings are presented for each period as follows. First, the main events and results are described. Then, the role played by library managers as leaders of change and how they dealt with different stakeholders are analysed (see Table 1). Finally, a comparison of the findings regarding leadership and stakeholder management is provided.

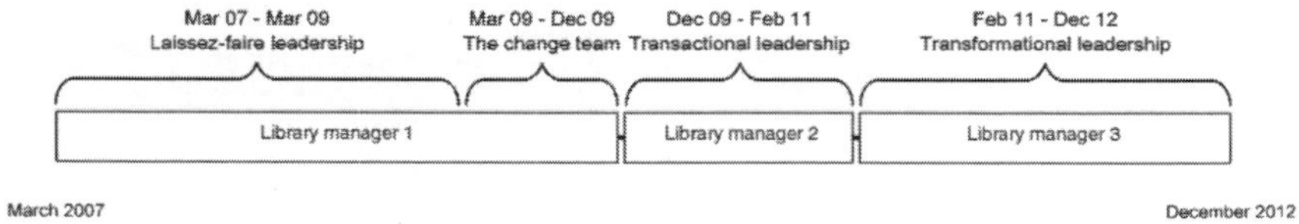

Figure 1. Overview of the periods of the library change process

4.1. First period: "Laissez-faire" leadership (March, 2007–March, 2009)

The change began in March 2007 when the ICP at the university level started to set up the library change as one of its sub-projects. A diagnostic of the library, on initiative of the ICP, was developed by an external group of international library experts, including a young and a senior chief librarian. Several recommendations were identified and communicated with the university's top administrators and librarians. Although several meetings and informal conversations took place among library staff and the ICP, the planning of change was not clearly and explicitly formulated. After almost two years, the interviewers recognized that the change had not been successfully implemented. This conclusion was corroborated by reports and observations. Organizational changes, such as a new library regulation and a new departmental structure, were not executed. A number of technological improvements were implemented by the ICT responsible, including: the installation of new computers, servers and storage equipment; the creation and implementation of a new Website; and higher-speed access to Internet and Intranet. After several discussions, the need to retake efforts to boost the change process in the library was recognized by top management. See **Sucozhañay et al. (2011)** for a detailed description of this period.

Table 1. Summary of findings.

Period/ Library managers	Transformational Leadership*	Stakeholder management **	Changes implemented
Laissez-faire leadership Library manager 1	No transformational behaviours exhibit.	Hardly stakeholder management practices performed.	Technological changes****
The change team Library manager 1 + Change Team***	No transformational behaviours exhibit.	Library Council integration, only *structured attempt* to involve stakeholders.	Technological changes****
Transactional leadership Library manager 2	No transformational behaviours exhibit.	Stakeholder management was not systematic, nor proactive. Good relationships with specific stakeholders.	Technological changes
Transformational leadership Library manager 3	Transformational behaviour exhibit.	Stakeholder management was not systematic, nor proactive. Good relationships with specific stakeholders.	Technological and organizational changes

* Transformational leadership behaviours: individualized consideration, inspirational motivation, intellectual stimulation, idealized influence.

** Stakeholder management practices: identification, analysis, formulation of strategies; facilitation of the process to implement these strategies and evaluation of strategy implementation.

*** The results are referred to LCT. Results for the library manager are similar to the first period.

**** Technological changes were in charge of the ICT responsible.

Regarding the leadership style, the library manager who was expected to take the lead in implementing the change, was perceived as playing a rather low-profile role and not exhibiting transformational leadership behaviours. Librarians saw him/her as someone who was highly experienced and trained but who failed to provide them with regular feedback. The library manager's leadership style was characterized as allowing librarians great freedom to propose and execute activities while avoiding getting involved and making decisions. An interesting fact, regarding the leadership on this period, is the little clarity on who was leading the change. For instance, on the one hand, the technological changes were conducted by the ICP and ICT responsible. On the other hand, when the library manager was asked about the leadership, he/she identified the ICP as leader of the change. Likewise, the Vice Rector, as the highest authority of the library, was also identified as responsible for the change. Unfortunately, the vision of change of these different actors was not aligned. The library manager explained that his/her idea of change was oriented to solve librarianship issues, such as cataloguing and processing, while the Vice Rector and ICP in coalition with the ICT responsible, were mainly focused on technological improvements. This disconnection contributed to the isolation of the library manager.

In addition to this lack of alignment, stakeholder management practices were not observed or mentioned in the interviews. The relationship between the library manager and stakeholders was weak; consequently, neither influenced the librarians, nor university's top administrators, and other actors to align forces for change (as shown in Figure 2a).

4.2. Second period: The change team (March, 2009–December, 2009)

After the first attempt to forge the change into the library, a new strategy was planned. Thus, an external expert updated the diagnosis of the library and drew the library change path but now with the support of a local Library Change Team (LCT) and a Library Council. The LCT was formed by an organizational consultant, a thesis student, an internship student of a European university, and the ICT responsible. The Library Council consisted of representatives from academic and library staff, and university top and middle management, formally appointed by the Vice Rector.

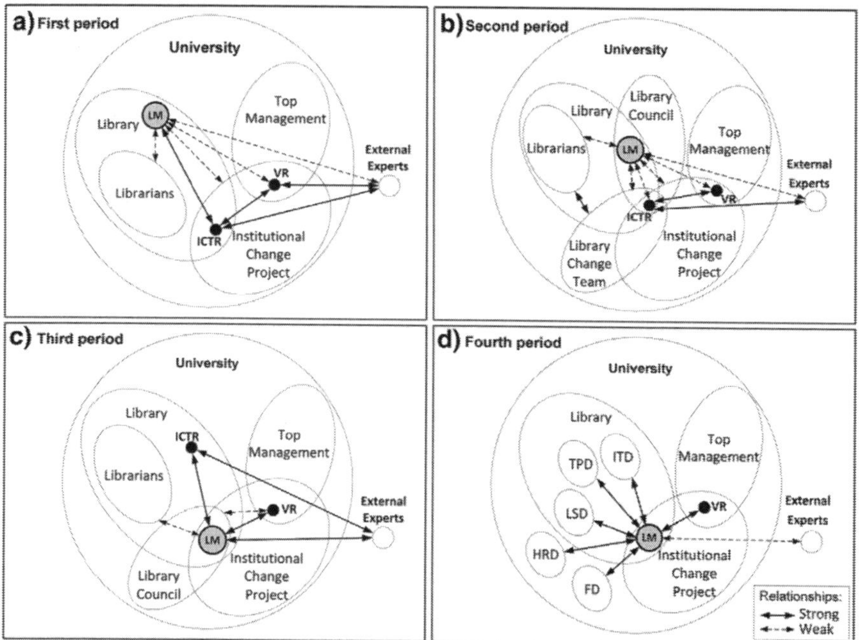

Note. – LM = Library Manager. VR = Vice Rector. ICTR = Information and Communication Technology responsible. HRD = Human Resource Department. FD = Financial Department. ITD = Information Technology Department. TPD = Technical Process Department. LSD = Library Services Department.

Figure 2. Library managers' relationships with key stakeholders during the change attempts.

The diagnosis showed that the library situation had not changed, except for some technological improvements. After several meetings with the participation of the library staff, LCT, Library Council and the external expert, a general guide for the library change, including vision, mission and five main objectives, was set out and validated by university's top administrators. The objectives were: 1) adapting the organizational structure, 2) optimizing the internal processes, 3) improving the interpersonal relationships, 4) positioning of the library image, and 5) updating the library regulations. The LCT took charge of the implementation process. Important to note is that the library manager was not an active actor of the LCT, due to, as mentioned above, his/her misalignment of the change vision.

During the next six months, the activities were geared to implement this plan. The technological change, led by the new ICT responsible, continued. A new library system and a new cataloguing format were implemented. However, the gap between the implemented technology and ideas, and knowledge and skills of the library staff remained wide. As several librarians explain, many technologies and services were implemented without clearly knowing the theory and philosophy behind them. For example, Facebook services were implemented without a clear idea of its importance as part of Libraries 2.0. More details on this period can be found in **Sucozhañay et al. (2011)**.

Concerning the leadership of this period, no change in the library manager performance was found. The LCT had a prominent role in this period. This team had a direct communication with the Vice Rector and librarians. Nevertheless, transformational behaviours of its members were not found. The limitation of the LCT was its little political power to take decisions, since none of its members had a formal position at the university.

Regarding stakeholder management practices, the relationship between the library manager and the stakeholders remained relatively poor (see Figure 2b). No systematic stakeholder management practices were identified. The establishment of the Library Council could be considered as a *"first structured attempt"* to involve several stakeholders though this initiative was boosted by the ICP. The LCT carried out the establishment of the Library Council. Unfortunately, university's top administrators and the library manager hardly participated in the

Library Council, thereby minimizing the potential stakeholder management.

4.3. Third period: The managerial leader (December 2009–January 2011)

In December 2009, the university's top administrators appointed the ICT responsible as library manager. The good relationship and shared understanding of the content of change allowed direct communication between the library manager, ICP members, and the Vice Rector, thereby eliminating previous disconnections. Activities in this period were focused on the implementation of proposals that were developed in the previous period, such as the new library regulation, training of library staff, and dissemination of online services. Despite the emphasis placed on the development of proposals for a new organizational structure, including departmentalization schemes, process optimization, and manual functions, none of these contributions were implemented. Librarians and Library Council members perceived that no big organizational changes were achieved. The technological change continued to be an important element in the library change. As a result, a new online public access catalogue was implemented, and the library Web portal was redesigned to incorporate Library 2.0 tools, such as blogs, social networking, messaging applications and Really Simple Syndication (RSS) feeds.

Regarding leadership behaviours, the majority of librarians perceived the new library manager as more active, communicative, task-oriented, and having a high commitment with the change process. The library manager put emphasis on setting short-term goals, rules and procedures. Despite these favourable characteristics, the new library manager was not perceived by librarians as having enough influence, librarian knowledge, and expertise to motivate and inspire followers. The relationship between the new library manager and librarians tended to be transitory and not based on emotional bonds. Accordingly, transformational behaviours were almost absent. Interestingly, although librarians recognized the need for change, it did not necessarily mean radical change, and they often felt threatened by the new technological innovations promoted by the library manager.

Stakeholder management in this period was not systematic but rather incident-based. For instance, the library manager directly worked together with university's top administrators to achieve specific results such as the final approval of the new library regulation. The Library Council continued having meetings and discussions on how to implement the change and how to get enough resources for it. Additionally, they provided the political support and feedback for the process of approval of the new library regulation. Interestingly, after almost one year, the work intensity of the Library Council decreased due to both few meeting requests and the intermittent attendance of their members. The relationships of key stakeholders in this period are presented in Figure 2c.

The analysis of this period prompts the conclusion that although the new library manager had a good stakeholder relationship, he/she did not display transformational leadership qualities. This situation limited the library manager's ability to influence stakeholders to achieve the library vision.

4.4. Fourth period: The transformational leader (February 2011–December 2012)

At the end of January 2011, the second library manager resigned to accept another position. As a consequence, an experienced librarian was appointed as new library manager by university's top administrators.

In this period, the new library manager was able to implement the new organizational structure that had not yet been implemented, due to low participation and commitment rates of librarians. The new organizational structure consisted of one head and three departmental units: Technical Process, Library Services, and Information Technology, each with their respective coordinator. After its implementation in April 2011, librarians noted that the new structure allowed for the specialization, promotion, teamwork, improvement of coordination, and monitoring of library functions.

With the new structure implemented and running, efforts were made to develop the library strategic plan for 2012–2017. To give support in this task, a new external expert visited the library in May 2011. After a week of work, the library staff and university's top administrators agreed on the strategic lines. Librarians recognized the importance of the plan in clarifying the library direction for the upcoming years. The strategic lines were: 1) organization and co-operation, 2) support of research, 3) support of teaching and learning, 4) digital library, 5) staff capacity building, and 6) quality management.

The emphasis in this period was then put on the implementation of the library strategic plan. Digital services, such as a digital repository and remote access to bibliographic databases, were implemented. Library staff and users were trained in topics related with digital services. Additionally, the library participated actively in the university accreditation and improved its cooperation and networking with other libraries.

The new library manager was recognized by all librarians as leader of the change process and displaying transformational leadership qualities. Attributes, such as communicative, persistence, envisioning future, commitment to goals, valid knowledge and skills, ideas generator, team building among others, were used to describe the library manager's leadership style. Attitudes that, according to the literature, fall into the four transformational behaviours: idealized influence, inspirational motivation, intellectual stimulation, and individualized consideration (**Bass & Riggio, 2006**). The following quotation illustrates this view: "*The library manager is leading the process of change, and has become a* *leader because of the determination, knowledge and passion about the library"* (Librarian).

Regarding the stakeholders, the library manager recognized the importance of the Library Council because its members are important stakeholders; however, no effort was made to reactivate it. Instead, to implement certain activities, the library manager initiated direct communication with university's top administrators and different university departments and units, such as Human Resources and the Financial Department (see Figure 2d). Although a good relationship with these stakeholders could be seen, neither a systematic nor a proactive management of the stakeholders could be identified. An overview of the findings implemented in all investigated periods, regarding transformational leadership, stakeholder management, and main changes, is given in Table 1.

5. DISCUSSION

The transition from a traditional library to a Library 2.0 in Latin America should be seen as an integral and complex change. As can be seen in this case study, implementing these types of changes requires a good deal of vision, time, and effort (**Jantz, 2012**). Technological and organizational components have to be managed simultaneously; otherwise, this shift of paradigm must settle for merely automating traditional library services. Accordingly, **Graetz and Smith (2009)** argue that many libraries continue to uphold traditional structures while they try to implement new approaches. In fact, the key to successful implementation of any new model is usually not the technology but human factors (**Williamson, 2008**), and this case study is not the exception. Evidence shows that people are essential to the success of a change process, but they often resist (**Marshall, 2008**). Zimmerman (2006) posits that resistance is a major factor in the failure of change. Therefore, library managers have to develop a participative process that encourages librarians and other stakeholders to work together to achieve a common goal, while preventing significant resistance to change (**De Bruyn & Kruger, 2001**). In this case, change resistance was not highly significant, it was mainly related to the way that the change was led.

The literature on leadership styles and stakeholder management provides insight into the role of library managers in the implementation of change. The current findings reinforce the conclusion that transformational behaviours contribute to positive outcomes during times of change in academic library settings (**Albritton, 1998**). Transformational leadership style stresses the emotional and symbolic aspects of change (**Yukl, 2013**). For instance, leaders usually use emotional support to prompt followers' commitment, and thus reduce resistance. **Stringer (2002)** posits that leadership practices cannot be measured as 'good-bad' issues but rather as 'effective-less effective' issues. In this case, the third library manager, who displayed higher transformational leadership behaviours, was able to inspire, motivate, and facilitate change not only in technological aspects but also in organizational aspects. A transformational leader motivated library staff by means of communicating an appealing vision, demonstrating high levels of trust, persistence and knowledge about the library issues, attention to individualized development, and the ability and willingness to provide intellectual stimulation (**Bass, 1990**, p. 30). The second library manager was recognized as task-oriented and less effective in achieving an integral change. This insight further supports the idea that although transactional leadership style is appropriate in some settings, only transformational leadership can motivate the actions needed to achieve successful change (**Kotter, 1996**).

Additionally, during the whole process, only one of the three library managers was perceived as a transformational leader. This drives our attention to the way library managers were appointed. It was found that no analysis of leadership skills was performed before the appointment of library managers to their positions. This finding is in agreement with **Sukram and Hoskins' (2012)**findings, which show that library managers are often appointed without considering their leadership skills. In this study, it is therefore argued that one "Achilles' heel" of the library change was to assume that appointed library managers would be able to serve as transformational leaders. Traditionally, the selection of library managers is the exclusive responsibility of the university's top administrators. Unfortunately, these authorities lack time and experience to assess the candidate's leadership qualities.

Furthermore, as mentioned in the literature review, academic libraries are usually part of a parent organization, such as a university. This makes library change seen as a phenomenon that occurs in interaction with a wider range of relevant stakeholders (**López & Vargas, 2012**; **Solís & López, 2000**). For instance, in this case study, strategic decisions and resource allocation depended not only on library managers' decisions but also on university top administrators and other stakeholders, such as the finance director and deans' decisions (**Budd, 2005**; **López Segrera, 2008**; **Torres & Schugurensky, 2002**; **vanDuinkerken & Mosley, 2011**). This fact drives the discussion of the stakeholders' value in developing a major change in library settings, which has been corroborated in previous studies, such as **vanDuinkerken and Mosley (2011)**. According to**Turner (2007)** and **Kotter (1995)**, stakeholders' attitudes, including resistance or willingness to change, depend not only on the content of the change but largely on how the transition is led. Nevertheless, most of the library leadership literature has focused on the leader-subordinate relationship (**Tam & Robertson, 2002**; **Williamson, 2008**).

The library manager, as change leader, has to engage stakeholders with the goals of the change (**Stavridis & Tsimpoglou, 2012**; **vanDuinkerken & Mosley, 2011**). However, as illustrated in Figure 2, findings show that most of the time, library managers have a rather limited and weak relation with stakeholders (**vanDuinkerken & Mosley, 2011**). **Horn (2008)** posits that each stakeholder may require a different approach to be managed. For instance, library managers have to be aware of the strong contribution that university top management can make to achieve radical change. Nevertheless, even though support and commitment to change is expected and offered from university's top administrators, it fluctuates during the change process. Low attention and engagement are sometimes evidenced. Therefore, library managers have to work on building and strengthening their relationship with university top management to increase the chances to succeed. Likewise, university's top administrators, as principals responsible for library development, should take sound decisions to avoid bottlenecks during the implementation of change, such as providing budget and resources.

Stakeholder management practices were barely identified in our case study. Nevertheless, it does not necessarily mean that library managers developed different relationship approaches with their stakeholders. Remarkable is that, although all three library managers recognized the importance of university actors, none of them performed a systematic practice to identify, analyse, and manage these stakeholders. This finding reinforces the reported experience on previous studies that leaders often fail to identify, analyse, and manage stakeholders (**Savage et al., 1991**). From a pragmatic point of view, the importance of stakeholder management must be recognized; managing highly diverse stakeholder groups effectively, is not an easy task as demonstrated in this case study. There were some isolated practices that library managers performed in order to negotiate and engage some stakeholders. For instance, the "Library Council" experience was an interesting attempt to incorporate other stakeholders. Nevertheless, even with a promising start and recognition as part of the organizational structure of the library, the committee stopped its work by the end of the third period. Although this experience would have been an interesting way to integrate stakeholders, it had its limitations. First, the selection of Library Council members was guided by a criterion of "experience in research" but not according to their power, legitimacy, and urgency (**Mitchell, Angle, & Wood, 1997**); three key stakeholder attributes that have received reasonable empirical support to identify and analyse "who really counts" when performing stakeholder mapping and evaluation (**Laplume, Sonpar, & Litz, 2008**; **Mitchell, Angle, & Wood, 1997**). Second, no analysis or management strategies of these actors was generated by leaders. For instance, most of the times, key stakeholders were very busy with the day-to-day management of their own faculties or departments, simply lacking time and energy to invest in collaboration with other parties. Finally, Library Council recognition in the third period was still low, limiting its capacity to influence decision-making processes.

Several methodological proposals on how leaders can manage stakeholders can be found in the literature (**Mitchell, Angle, & Wood, 1997**; **Savage et al., 1991**;**Svendsen, 1998**). However, it is not clear why leaders do not put any of these recommendations into practice. One possible explanation, coming from this study, is that library managers usually see the university top administrators and the administrative and academic directors as individuals to whom they should report, rather than actors to be influenced. Another possible explanation is the lack of time library managers and stakeholders have for such activity. Based on the case study analysis, the general conclusion is that although the importance of stakeholders is recognized by library managers, their management is more reactive than proactive. This prompts the need of library leaders to understand more deeply the stakeholder management approach, which means to recognize the stakeholders, their motivations, and interests. In addition, to increase the likelihood of a successful change, an important activity is the formulation of strategies, and subsequently the facilitation of processes to implement such strategies (**Mitchell, Angle, & Wood, 1997**; **Savage et al., 1991**).

5.1. Practical implications

In general, our findings suggest that library managers, as change leaders, should perform transformational behaviours and actively manage the stakeholders in order to achieve a successful change. This double approach – transformational leadership and stakeholder management – used to analyse the role of library managers, recognizes the complexity of relationships in an academic library change implementation.

Sukram and Hoskins (2012) recommend that library managers be trained on leadership competencies and management of the stakeholders. Nevertheless, managers are rarely trained in these issues (**Holm-Nielsen, Thorn, Brunner, & Balán, 2005**; **vanDuinkerken & Mosley, 2011**). Library managers have to look for transformational leadership skills to encourage organizational change. **Bass (1990)** posits that transformational leadership skills can be learned, and plenty of methodologies to implement stakeholder management can also be found in literature. However, as these concepts have been developed mainly in business arenas, library managers have to adapt them for the academic field. Training in both transformational leadership and stakeholder management will help library managers improve their performance as change leaders.

5.2. Research limitations

A limitation of this study is that data collection for the analysis about leadership and stakeholder management was conducted through retrospective interviews, where interviewers may have omitted or filtered details. Nevertheless, data triangulation and data collection at different moments allowed a reduction of this risk. An additional limitation is that this study was focused on the role of leaders in involving library staff and internal stakeholders within the university. External stakeholders such as suppliers, other libraries and government institutions were not analysed. Finally, the last limitation is that the change process analysis was performed until December 2012; however, the change process is still in progress.

6. CONCLUSIONS

Transforming a traditional library into a Library 2.0 has been investigated widely, often with a strong emphasis on the technological aspects. This paper illustrates the importance of the "human factor" in technological innovations. It was found that transformational leadership and stakeholder management play a critical role. To ensure the success of change, leaders have to align the vision and the implementation strategies of change among stakeholders at multiple levels. In this particular case, although library managers recognized the importance of different stakeholders, findings show that there is no specific strategy to manage them. In fact, stakeholder management actions in this case study were reactive and unsystematic. Library managers should act as transformational leaders creating sustainable and trustful relationships not only with the library staff but also with other stakeholders to reach this goal.

ACKNOWLEDGMENTS

We wish to thank the staff of the Regional Documentation Centre at University of Cuenca for their cooperation throughout this study. Special thanks to the Flemish Interuniversity Council (VLIR), the University of Cuenca, and the National Secretariat of Higher Education, Science, Technology and Innovation of Ecuador (SENESCYT) for supporting this research project.

REFERENCES

1. Albritton, R.L. (1998). A new paradigm of leader effectiveness for academic libraries: An empirical study of the Bass (1985) model of transformational leadership. In T. Mech & G. B. McCabe (Eds.),

Leadership and academic librarians (pp. 66–82). West-port: Greenwood Press

2. Antonakis, J., Avolio, B.J., & Sivasubramaniam, N. (2003). Context and leadership: an examination of the nine-factor full-range leadership theory using the Multifactor Leadership Questionnaire. The Leadership Quarterly, 14(3), 261–295. doi: 10.1016/S1048-9843(03)00030-4

3. Armenakis, A.A., & Bedeian, A.G. (1999). Organizational change: A review of theory and research in the 1990s. Journal of Management, 25(3), 293–315. doi: 10.1177/014920639902500303. Retrieved July 16, 2014, from http://www.scribd.com/ doc/ 189456956/Organizational-Change-A-Review-of-Theory-and-Research-in-the-1990s

4. Avolio, B.J., & Bass, B.M. (1991). The full range of leadership development: Basic and advanced manuals. Binghamton, NY: Bass, Avolio & Associates

5. Bass, B.M. (1990). From transactional to transformational leadership: Learning to share the vision. Organizational Dynamics, 18(3), 19–31. Retrieved July 16, 2014, from http:// strandtheory.org/images/From_transactional_to_transformational_-_Bass.pdf

6. Bass, B.M., & Riggio, R.E. (2006). Transformational leadership (2nd ed.). Mahwah, NJ: Lawrence Erlbaum

7. Bryson, J. (2004). What to do when stakeholders matter. Stakeholder identification and analysis techniques. Public Management Review, 6(1), 21–53. Retrieved July 16, 2014, from https://www.hhh.umn.edu/people/jmbryson/pdf/stakeholder_identification_analysis_techniques.pdf

8. Budd, J. (2005). The changing academic library: Operations, cultures, environments (Vol. 56). Chicago: Association of College and Research Libraries

9. Castiglione, J. (2006). Organizational learning and transformational leadership in the library environment. Library Management, 27(4/5), 289–299

10. Crosby, P.B. (1996). The absolutes of leadership. San Francisco: Jossey-Bass Publishers

11. Cummings, T.G., & Worley, C.G. (2005). Organization development and change (8th ed.). Mason, OH: Thomson/South-Western

12. De Bruyn, H., & Kruger, S. (2001). Strategic management: A southern African perspective (3 ed.). Johannesburg: Entrepro Publishers

13. Freeman, E., & McVea, J. (2001). A stakeholder approach to strategic management. Darden Business School Working Paper No. 01-02. Retrieved July 16, 2014, from http://papers. ssrn. com/ sol3/papers.cfm?abstract_id=263511

14. Freeman, R.E. (1984). Strategic management: a stakeholder approach. Boston, MA: Pitman

15. Govender, P., Moodley, S., & Brijball Parumasur, S. (2005). The management of change at selected higher education institutions: An exploration of the critical ingredients for effective change management. Journal of Industrial Psychology, 31(1), 78–88. Retrieved July 16, 2014, from http://www.sajip.co.za/index.php/ sajip/article/viewFile/186/183

16. Graetz, F., & Smith, A.C. (2009). Duality theory and organizing forms in change management. Journal of Change Management, 9(1), 9–25. doi:10.1080/14697010902727146. Retrieved August 22, 2014, from: http://www.aaronctsmith.com/Article%20PDFs/Graetz%20JCM%20 2009.pdf

17. Holm-Nielsen, L.B., Thorn, K., Brunner, J.J., & Balán, J. (2005). Regional and international challenges to higher education in Latin America. In H. De Wit, C. Jaramillo, J. Gacel-Ávila, & J. Knight (Eds.), Higher Education in Latin America. The international Dimension (pp. 39–69). Washington, DC: The World Bank. Retrieved July 16, 2014, from http://siteresources.worldbank.org/ EXTLACREGTOPEDUCATION/Resources/Higher_Ed_in_LAC_Intnal_Dimension.pdf

18. Horn, A. (2008). Strategic competence: to soar above. Library Management, 29(1/2), 5–17. doi: 10.1108/01435120810844603

19. Huang, Y.-C., & Liao, L.-C. (2011). A college leader's transformational leadership. Journal of Management Research, 3(2), 1–22. Retrieved July 16, 2014, from http://www.macrothink.org/journal/ index.php/jmr/article/view/646/483

20. Jantz, R.C. (2012). Innovation in academic libraries: An analysis of university librarians' perspectives. Library & Information Science Research, 34(1), 3–12. doi: 10.1016/j.lisr.2011.07.008. Retrieved July 16, 2014, from http://web.simmons.edu/~weigle/MGMT_INNOVATION/Jantz_InnovationAcademicLibraries.pdf

21. Jorgensen, D.L. (1989). Participant observation: A methodology for human studies (Vol. 15). Newbury Park, CA: Sage

22. Kotter, J.P. (1995). Leading change - Why transformation efforts fail. Harvard Business Review, 73(2), 59–67. Reprinted in Best of HBR (2006), Retrieved July 16, 2014, from https://secureweb.mcgill.ca/hr/sites/mcgill.ca.hr/files/leading_change_john_kotter.pdf

23. Kotter, J.P. (1996). Leading change. Boston, MA: Harvard Business School Press

24. Kotter, J.P., & Cohen, D.S. (2002). The heart of change: real-life stories of how people change their organizations. Boston, MA: Harvard Business School Press

25. Kouzes, J.M., & Posner, B.Z. (2012). The leadership challenge: How to make extraordinary things happen (5th ed.). San Francisco, CA: Jossey-Bass

26. Laplume, A.O., Sonpar, K., & Litz, R.A. (2008). Stakeholder theory: Reviewing a theory that moves us. Journal of Management, 34(6), 1152–1189. doi: 10.1177/0149206308324322

27. López, J.J., & Vargas, J.G. (2012). Ambigüedad organizacional en la planeación estratégica. Revista de Administração FACES Journal, 11(2), 44–67. Retrieved July 16, 2014, from http://www.fumec.br/revistas/facesp/article/view/1215/837

28. López Segrera, F. (2008). Tendencias de la educación superior en el mundo y en América Latina y el Caribe. Avaliação: Revista da Avaliação da Educação Superior (Campinas), 13, 267–291

29. Marshall, S. (2008). Leading change at the middle: Stories of higher education middle leaders 'success'. (Unpublished thesis submitted in partial fulfillment of the degree of Master of Educational Leadership and Management), Unitec Institute of Technology, New Zealand. Retrieved October 1, 2014, from http://hdl.handle.net/10652/1304

30. McGuigan, G.S. (2012). Addressing change in academic libraries: a review of classical organizational theory and implications for academic libraries. Library Philosophy and Practice 2012, paper 755. Retrieved July 16, 2014, from http://digitalcommons.unl. edu/cgi/viewcontent.cgi?article=1829&context=libphilprac

31. Miles, M.B., & Huberman, A.M. (1994). Qualitative data analysis: an expanded sourcebook. Los Angeles: Sage Publications

32. Mitchell, R.K., Agle, B.R., & Wood, D.J. (1997). Toward a theory of stakeholder identification and salience: Defining the principle of who and what really counts. Academy of Management Journal, 22(4), 853–886. Retrieved July 16, 2014, from http://courses.washington.edu/ilis580/readings/Mitchell_et_al_1997.pdf

33. Munatsi, R. (2010). Implementation of library 2.0 services in African academic and research libraries: need for fundamental rethink. In P. Jain, & R. Nfila (Eds.), Proceedings of the 19th Standing Conference of Eastern, Central, Southern Africa Library and Information Associations (SCESSAL), held at University of Botswana, Gaborone, 6–9 December 2010. Gaborone: Botswana Library Association

34. Riggs, D.E. (2001). The crisis and opportunities in library leadership. Journal of Library Administration, 32(3–4), 5–17. doi: 10.1300/J111v32n03_02

35. Savage, G.T., Nix, T.W., Whitehead, C.J., & Blair, J.D. (1991). Strategies for assessing and managing organizational stakeholders. The Executive, 5(2), 61–75

36. Solís, P., & López, B. (2000). El concepto de las anarquías organizadas en el análisis organizacional. Contaduría y Administración, 197, 35–45. Retrieved July 16, 2014, from http://www.ejournal.unam.mx/rca/197/RCA19704.pdf

37. Stavridis, S., & Tsimpoglou, F. (2012). EFQM in academic libraries: The application of a quality management model at the University of Cyprus library. Liber Quarterly, 22(1), 64–77. Retrieved July 16, 2014, from http://liber.library.uu.nl/index.php/lq/article/ view/ 8041/8413

38. Stephens, D., & Russell, K. (2004). Organizational development, leadership, change, and the future of libraries. Library Trends, 53(1), 238–257. Retrieved July 16, 2014, from http://cdigital.uv.mx/ bitstream/123456789/5984/2/Liderazgo%202.pdf

39. Stringer, R. (2002). Leadership and organizational climate: the cloud chamber effect. Upper Saddle River, NJ: Prentice Hall

40. Sucozhañay, D., Sigüenza-Guzman, L., Zhimnay, C., Cattrysse, D., Euwema, M., & De Witte, K. (2011). Change management strategies for successful implementation of a library 2.0. A case study in Ecuador. In L. Gómez Chova, I. Candel Torres, & A. López Martínez (Eds.), Proceedings of the 5th International Technology, Education and Development Conference (INTED 2011) (pp. 1051–1060). Valencia, Spain: International Association of Technology, Education and Development, IATED

41. Sukram, P., & Hoskins, R. (2012). Leading the way : the managerial skills of library managers in South African university libraries. Innovation, 45, 94–119

42. Svendsen, A. (1998). The stakeholder strategy: profiting from collaborative business relationships. San Francisco, CA: Berrett-Koehler Publishers

43. Swanborn, P. (2010). Case study research: What, why and how? Los Angeles: Sage

44. Tam, L., & Robertson, A. (2002). Managing change: libraries and information services in the digital age. Library Management,

23(8/9), 369–377. doi: 10.1108/01435120210439843. Retrieved July 16, 2014, from http://hub.hku.hk/bitstream/10722/ 48727/1/ 84942.pdf?accept=1

45. Torres, C.A., & Schugurensky, D. (2002). The political economy of higher education in the era of neoliberal globalization: Latin America in comparative perspective. Higher Education, 43(4), 429–455. doi: 10.1023/a:1015292413037. Retrieved July 16, 2014, from http://www.del.ufrj.br/~ricardo/DocsArcuSur/Carlos.pdf

46. Turner, D.-M. (2007). Clinician readiness for transition to a fully integrated electronic health care delivery system. (PhD. dissertation), Walden University, Minnesota. Available from Dissertations & Theses: A&I. (Publication No. AAT 3247595)

47. vanDuinkerken, W., & Mosley, P.A. (2011). The challenge of library management: Leading with emotional engagement. Chicago: American Library Association

48. Verschueren, M. (2014). Leadership styles of head nurses and quality of care. (PhD), University of Leuven, Leuven

49. Williamson, V. (2008). Relationships and engagement: The challenges and opportunities for effective leadership and change management in a Canadian research library. Library Management, 29(1/2), 29–40. doi: 10.1108/01435120810844621

50. Yin, R. (2009). Case study research. Design and methods (4th ed. Vol. 5). California: Sage

51. Yukl, G. (1999). An evaluation of conceptual weaknesses in transformational and charismatic leadership theories. The Leadership Quarterly, 10(2), 285–305. doi: 10.1016/S1048-9843(99)00013-2. Retrieved July 16, 2014, from http:// www. researchgate.net/publication/223878512_An_evaluation_of_conce ptual_weaknesses_in_transformational_and_charismatic_leadershi p_theories/file/72e7e52ed0e2757691.pdf

52. Yukl, G.A. (2013). Leadership in organizations (8th ed.). England: Pearson Education, Limited

APPENDIX 1: SEMI-STRUCTURAL INTERVIEW GUIDE TO STAKEHOLDERS

1. Description of the change process

1.1 In general, what did the change process consist of?

1.2 What was the problem or goal that was intended to solve or achieve?

1.3 Could you describe how the change process (planning, implementation, assessment, etc.) was held?

1.4 What were or are your interests / aspirations / concerns on this change? Did you agree with the change?

1.5 How did you participate in this process?

2. Analysis of stakeholder management

2.1 How did you get involved in this process? Who contacted you? What was/is the role that you played/play in the process?

2.2 Why do you think you were involved in this process? At what point of the process you got involved?

2.3 What other stakeholders were involved? Could you enlist and analyse them in terms of their role in the process and their power to influence it?

2.4 Do you know how these actors were involved? Why were they involved?

2.4.1 Do you know if there was a stakeholders' analysis to decide their involvement? What were the criteria used for such analysis?

2.5 How was the relationship between the actors? Was there any conflict?

3. Leadership

3.1 Who exercised leadership in this process of change? Why do you consider him or her as a leader?

3.2 What behaviour did the leader exhibit? What activities did the leader perform? How did the leader make his/her decisions?

3.3 How was your relationship with the leader? How was the leader's relationship with the other actors?

3.4 What type of communication was used between the leader and the other stakeholders?

3.5 From your perspective, how did leadership influence the change process and the results achieved?

4. Change process results

4.1 What were the results achieved? Why? What results were not achieved? Who were the key stakeholders to achieve these results? Why?

4.2 What do you think were the main problems / bottlenecks / resistance occurred in this process?

4.3 How did the organizational structure of the university influence the implementation of change?

4.4 What recommendations would you make to improve the management of change in universities?

APPENDIX 2: LIST OF CODES USED FOR THE ANALYSIS

Category	Codes	Sub-codes
Change Process (CP)	CP_ Change Context (CC)	CC_University context CC_Library context
	CP_Time line CP_Planning CP_Objectives CP_Implementation CP_Results CP_Evaluation	CP_First Period* CP_Second Period* CP_Third Period* CP_Fourth Period*
	CP_Responses to change (RC)	RC_Resistance RC_Willingness RC_Ambivalence
	CP_Setbacks*	
	CP_Hits*	
Leadership (L)	L_Leaders identified	L_Leader period 1–2* L_Leader period 3* L_Leader period 4* L_University's top administrators * L_Institutional Change Project*
	L_Leaders behaviours	
	L_Expected role of leaders*	
Transformational leadership (TL)	TL_Behaviours	TL_Intellectual stimulation TL_Idealized influence TL_Inspirational motivation TL_Individualized consideration
	TL_Effectiveness	
Stakeholder management (SM)	SM_Identification of stakeholder	SM: Identification strategies SM: Stakeholders identified* SM: Characteristics of stakeholders*
	SM_Analysis of stakeholders	SM_Criteria used SM_Techniques used
	SM: Formulation and implementation of strategies	SM_strategies planned SM_strategies implemented
	SM: Evaluation of stakeholder management	
	SM_Other practices to manage stakeholders*	
	SM_Library Council*	
Leadership (L) + Stakeholder Management (SM)	L+SM: Relationship of the leader with stakeholders*	TL+SM: Relationship of the leader with stakeholders*

* Codes that emerged from data.

APPENDIX 3: EXAMPLE OF SUMMARIZING MATRIX: DISPLAY OF ACTIVITIES, OBJECTIVES AND OUTCOMES OF THE FIRST PERIOD.

<table>
<tr><th>Period</th><th>Main activities</th><th>Objectives</th><th>Outcomes</th></tr>
<tr><td rowspan="3">March 2007 – March 2009</td><td rowspan="3">Diagnosis of the library carried out by two international experts.

Visits to libraries at national and international level.

Development and implementation of the plan of change called "Project for technological improvement".

Development of a first draft of new library regulation and a departmentalization proposal.

Purchase and installation of new technological equipment.</td><td>1. Technological improvements.</td><td>a) Improved Internet access through the increase of the number of computers;
b) Training room in the Central library was built;
c) Improved library network (bandwidth);
d) Implementation of a unique web site for the library;
e) Installation of a closed-circuit security in the Central library;
f) New servers and storage equipment were installed.</td></tr>
<tr><td>2. Staff and users training.</td><td>Library staff and users attended several workshops.</td></tr>
<tr><td>3. Organizational objectives.</td><td>The organizational outcomes were mainly focused on the elaboration of the library regulation and a departmentalization proposal.

Departmentalization proposal was not operationalized.</td></tr>
</table>

Chapter 3

Drivers and Directions of Academic Library Development 2005

Michael Cotta-Schönberg

ABSTRACT

The purpose of this paper is not to guess what the future of academic libraries will be, but to identify and understand the present driving forces with a view to choosing and influencing directions of development, which will be advantageous to our users. Four major players will be considered: the ICT-industry, the academic institutions, the publishing industry, and the libraries.

1. INTRODUCTION

The purpose of this paper is not to guess what the future of academic libraries will be, but to identify and understand the present driving forces with a view to choosing and influencing directions of development, which will be advantageous to our users. Four major players will be considered: the ICT-industry, the academic institutions, the publishing industry, and the libraries.

2. THE ICT-INDUSTRY

It is evident that developments in technologies of information and communication technologies are a very potent driver of the development of academic libraries. Just think of the importance of clay tablets and papyrus for the development of libraries in the classical age and of Gutenberg's invention of movable types for the development of the modern library. What has surprised many has been the alacrity with which libraries have embraced the new ICT. In many universities the library has been considered to be a pioneer in the application of these technologies. It is evident, however, that the library has no influence whatsoever on the development of these technologies. Therefore, the challenge for libraries is to use these technologies in appropriate ways, and so to speak surf on the ICT-wave - preferably ahead of our users!

3. THE INTERNET AND THE WEB

Are the ICT-development and especially the internet as well as the recent personal communication technologies a direct threat to libraries? They may very well be. Those of you, who have seen the movie I robot may remember the desperate accusation of the presumed scoundrel, who cries out - just before he dies: "You would have banned the internet to save the libraries". And a daily paper in Copenhagen recently featured a headline saying: "The librarian of the future is Google". So, even if librarians themselves to do not see a possible conflict, others do.

The recent announcement of Google's project to store millions of pre-copyright books and to make them available on the internet is seen by many as the writing on the wall for libraries. For business libraries as a particular species of libraries this development is not important, though, since business studies to not greatly depend on books from the pre-copyright period. Still, the trend as such is highly significant and for other types of libraries, i.e. great humanities libraries with old collections, it may be very important indeed.

3.1 Search engines

Everybody is now aware that the large search engines, like Google, are extremely good for certain types of information searching of a factual nature, but not very good for other since they simply retrieve too many records at the same time as omitting many relevant records actually retrievable through the net. However, the development

of specialized search engines, for example Google Scholar, offers at least two awesome alternatives to traditional library services. Firstly, it seems that they may eventually provide the possibility for integrated catalogue searches among pluralities of libraries combined with localisation facilities directed to particular libraries. Secondly, Google Scholar is already covering a number of publishers of electronic journals, e.g. Blackwell UP, and making the full text available to users from those institutions, for example universities, which subscribe to these periodicals.

I recently searched Google Scholar for literature on supplier relations in connection with product development. One of the first items presented was a highly relevant article from a Blackwell e-journal. I clicked on the item and was immediately transferred to the system of Blackwell UP

which recognized me as a user from the Copenhagen Business School (CBS). As the CBS subscribes to Blackwell's e-journals, the system then allowed me to print or download the article directly. So, Google

Scholar transferred me to Blackwell UP's web page where the full text article was made directly available to me - bypassing completely the library system of the CBS.

As our users today are directly tuned into the great search engines and only enter the library systems as a second or third choice, this development threatens to render obsolete the library's search facilities for searching its own materials and - as far as e-materials is concerned - also for retrieving them. This challenge is one of the most serious to libraries today. Shall we fight on to be alternatives to the specialised search engines, or should we rather join the flow and systematically make our resources available through search engines?

4. THE ACADEMIC INSTITUTIONS

4.1 E-learning

At the Copenhagen Business School the concept of e-learning is a somewhat problematical one. The pedagogical vision of the CBS considers as essential the direct interfacing between teacher and students and between students themselves. E-learning is therefore taken to mean firstly a form of teaching to be included in the ordinary course at the discretion of teachers but not to replace face-to-face teaching. Secondly, it has come to designate e-based support of ordinary and distance teaching, for example in the form of student discussion facilities and the electronic provision of course materials.

Many libraries have developed e-learning facilities in connection with their courses in library instruction which may make them useful partners for their academic institutions in the e-learning development in general. However, there is a much more important task for libraries in the development of dynamic collections of e-based materials for individual university courses. Such collections will become quite essential in the future and replace the traditional printed course packs.

4.2 Scholarly publishing

The present scholarly publishing system has proven very resistant to change. Academic merit still largely depends on publications in high-profile international or US print journals, and there seems to be a perception on the part of university leaders that quality and therefore prestige is bound up with being published in paper in the traditional way. There are important forces counteracting this trend, of course, such as last year's report from a British parliamentary committee on scholarly publishing.

The present system is too costly and cannot cope with the ever-growing demand for academic publishing outlets. Young scholars are happy with e-journals and are aware that peer reviews and all that can perfectly well be maintained in connection with an e-publishing system. Finally, a growing number of important universities are realizing that there may be an alternative to the present system based on universities collaborating in publishing scholarly e-journals and not handing over their research product to publishers. Academic publishing may eventually develop into storing texts on the universities' own servers (open archives) and making them available through branded and quality assured interfaces assuming the format and the functions of the traditional journals.

In the long run, this trend presents a relief for the financial situation of libraries since it will alleviate the cost of periodicals. Universities may of course prefer to confiscate the savings for other purposes, for example for scholarly publishing! Moreover, it offers an opportunity for libraries to provide a new support service for universities in the form of open archives. Some libraries may even become central players in the academic publishing process itself, but it is doubtful that this will be a general trend. Academics will prefer to perform that task themselves.

4.3 Dissemination of knowledge to society

There is a growing trend for societies to expand the mission of universities. Besides the traditional functions of research and teaching the third major task of universities is now seen as dissemination of knowledge to society, including the general public. Academics themselves will of course carry the main responsibility for this work of dissemination, but they will need instruments and service structures to support it. One type is the expert base facilitating the access of the public to individual researchers at the university. Another is the research bulletin couched in popular language. A third is the access interface to the publications of researchers. In this area there is clearly an opportunity for libraries to provide a new service to the university.

4.4 Study environments

Some may have thought, that e-learning and distance learning make the physical university superfluous as a 'place' - in the strict sense of the word - for study. Why should students come to the university in order to study when they can perfectly well do that at home? The general trend seems, however, to be the exact opposite. It appears that intensified ICT-based communication does not diminish the need for young people to be together, directly and physically. A Danish university professor has recently proclaimed that the society of the future will not be the digital or the electronic society, but the convivial society.

In the same way, university students come flocking to the physical premises of the universities not only to participate in courses, but also to have study groups, to socialize, and ... to study in libraries. I believe that this trend is permanent: ICT does not abolish the social nature of man, but reinforces it. The trend is therefore an opportunity for libraries in the sense that it underscores the usefulness of one completely traditional library service and ensures an active future for it, independent of the print or electronic format of its collections.

4.5 Funding

Whether there will any general trend concerning the funding of university libraries is difficult to say. However, it may safely be predicted that a great number of libraries will experience a reduction of financial resources over the coming years. One reason will be that the growth of the web-based global information system tends to make the local library less visible and therefore - university managers may think - less necessary. For those libraries financing will be a serious challenge. Keeping the print-based library alive for the next 10-20 years and at the same time developing and expanding the e-library will demand a heavy focus on the optimal use of resources and a radical breakaway from traditional patterns of library expenditure and organization.

5. THE PUBLISHING INDUSTRY

5.1 E-resources

The most fundamental driver of academic library development is arguably the strategies of publishers concerning e-journals and e-books. The development of e-journals has definitely not yet run its full course, but it has now developed so far that the end of the print journal as an important information medium for academic libraries is in sight.

The next step is e-books. We shall have to assume that at some point all scholarly and professional literature will be stored electronically, with a number of access options ranging from direct electronic access to print-based access. Exactly where in the future this point may be located is really not important as long as we recognize the trend. What is important is rather to know when the professional and scholarly e-book will cease to be a marginal phenomenon and instead become a major component of our services.

Conventional wisdom in the 1990's was that it would probably take another generation. However, there are signs that major developments may be much closer to us. Among these signs are recent reports from the publishing industry that e-fiction literature now is beginning to take off, which means that publishers will not have to develop special e-production and distribution facilities for scholarly literature, but can simply apply methods developed for the economically much more important fiction area to the scholarly publishing area. This trend is hugely important to libraries. The academic library as we know it, the traditional library, stands and falls with the printed book.

5.2 Business model

As you are aware, the development of e-book services in libraries has until now been hampered by the fact that publishers have not as yet come up with a business model, which makes sense in academic library terms. That will change! A business model for e-books which makes sense to our users - but not necessarily to libraries - might be based on three elements:

1. Many but cheap transactions

Quite naturally publishers want to keep and if possible to expand their earnings and profits. As we have seen with e-periodicals they try - and some very successfully - to do this through a pricing strategy based on expensive and therefore relatively few usage transactions. I would recommend the opposite strategy: a pricing strategy based on many and very cheap usage transactions. And when I say cheap, I mean really, really cheap - why not one-cent per download? I am convinced that publishers could earn much more money that way.

2. Common search facilities

Publishers would store their e-books on their own servers or on host servers, but instead of trying to force users to search for relevant literature in each publisher's own system they would have to establish a common, user-friendly search facility whether it be a publisher-based system like SCOPUS or a search engine like Google Scholar.

3. Direct end user access

Finally, the business model would comprise and facilitate direct end user access to their system - with no need for passing through local library systems. Whether publishers can truly get their act together and create such a business model, may be doubtful. Their present track record is not promising. But if and when they do establish a radically new business model based on direct user access, libraries will really have to fight to stay in business.

6. THE LIBRARIES

What internal forces drive academic library development?

6.1 Need to survive

For one thing, institutions just like people have their need to survive and this applies to libraries, too. Survival is not guaranteed of course, but the very will to survive is necessary for survival, and if libraries loose that drive they have lost their basic dynamism. The fact that libraries want to survive means that they are still alive and kicking and that they may therefore be worth it.

6.2 Sense of mission

But survival instincts are not enough. Happily, libraries today are driven by much more, that is a sense of mission in terms of service and users' needs. Over the last generation there has been a radical transformation of academic libraries. Formerly, focus was to a large extent on books and collections and on cataloguing rules and internal procedures. Today, there is a much clearer sense of a service mission, and this sense not only justifies our continued existence, but it also prepares us for radical transformation, since it is not the library as an institution which is important but the users and their developing needs.

6.3 Innovativeness

The third potent internal driver of academic library development is innovativeness and entrepreneurship, the urge to identify new opportunities, to exploit them and in some cases to create them. Innovativeness is, of course, largely dependent on the individuals working in academic libraries. However, the library's history, culture and structure may seriously hinder or promote innovativeness, and it is therefore very important to promote a culture of innovation and entrepreneurship in our libraries and to develop structures, which facilitate it. Without these three internal drivers of academic library development - need for survival, sense of service mission, and innovativeness - the development of future information services to academics will largely bypass libraries.

6.4 Directions

So, there is a number of external and internal drivers of academic library development. What can libraries meaningfully do with them, what directions should they take?

6.5 Dehybridization

A first direction has to do with the hybrid library, in the sense that libraries today must have both print and electronic collections. There is a conscious choice to make whether to pursue the hybrid library as a matter of principle, to continue as long as possible to have print collections, and to remain libraries in the traditional format or - on the other hand - to reduce progressively, systematically and continuously the print element and to aim at the electronic library only. This is a path of dehybridization - not in the sense of returning to the print-based library, but in the sense of leaving it all together.

Academic business libraries will have the possibility of being at the forefront of this development because of one factor: the very high rate of obsolescence of business literature. In business libraries most of the print books lent to students and teachers for that matter have been published within the last 10 years. This means more or less that they shall only need print collections about ten years after the publishers have made all their new books available in e-format. For other types of libraries the development in this area may reasonably be slower, but the direction is the same: towards the e-point.

6.6 Study environments

To provide an environment conducive to study and concentration is a traditional function of the library and it is still much in demand by students. At the Copenhagen Business School at the same time as the usage of printed books has started to decrease and the usage of electronic resources is developing at an amazing speed, the number of visitors to the library has been increasing annually and attained 1.2 million in 2005. The ordinary academic library of the future may not be a place for printed books, but it will certainly be a place for students and for study. In this sense, the library will continue to have a physical presence in the university providing a fulcrum for the future library organization.

6.7 Personal service

Users will be able to do a lot themselves, and for various reasons - optimization of resources and user satisfaction - libraries shall actively support self-service. However, the global information service available to users - including those information resources subscribed to or owned locally - will be a very complex one, and in spite of the development of specialised search engines, it will become more and more complex.

One of the important challenges for librarians will be to show that they can make a real difference in this respect. The job is to develop the librarian into a highly competent information specialist with a deep knowledge of the functioning of the global information system, with a specialist knowledge of the information resources relevant to the users of the particular institution, and with contact skills and pedagogical skills which translate the specialist competence into a function of advice and guidance useful and attractive to users. Moreover, this service has to be made available to users on the users' own terms, directly in the study environments provided by the library, and indirectly through other means, e.g. electronic means (chat, e-mail, web page, SMS).

6.8 The centrifugal library

The continued existence of the library as a physical place does not mean that the library as an organisation should be physically centred. To the contrary, it may become possible and necessary to have library staff working much closer to the user groups and offering personal services at decentralized service points or possibly through staff working directly with researchers in the institutes. For this to become reality, librarians will have to develop new professional identities and become institute persons rather than library persons, or they will have hybrid identities as library persons and institute persons.

6.9 Scholarly publishing and dissemination of knowledge

Following the lead of our American colleagues, European librarians may add a new function to their traditional ones by engaging in scholarly publishing through the maintenance of open archives and institutional repositories and there may be scope for supporting the publication of marginal journals.

Also, libraries may have an important role to play in the new functions of universities with respect to dissemination of knowledge to the public by developing and maintaining appropriate interfaces to their open archives, through current information on new publications from the university and its researchers, and through expert bases facilitating public access to researchers at universities and business schools.

It is clear, that libraries will in this area have to work together with the communication office of their university, whose function they are not qualified to perform and therefore should not try to take over. The specific role of the library will be to develop the instruments of dissemination and other activities based directly on their repository function for the university's and researchers' publications.

6.10 Information literacy

The young people do not arrive at the university without information literacy and skills. To the contrary, they have been using personal communication technologies, computers and the web since early childhood. Still, they lack - sometimes to an astonishing degree - information skills at an academic level. The ease of using search engines make them unaware how much relevant information they may be missing, and there is a tendency to use web-based information uncritically. There is, therefore, still scope for the library promoting information literacy at the academic level, either as separate courses or tutorials for students, or as an integrated part of the curriculum, for example in connection with the writing of student papers.

6.11 Search engines: if you cannot beat them, join them

Search engines are here to stay. Formerly, it was a widespread doctrine that libraries should teach users to use the library's own catalogue as the first step in information searching and only proceed to engine searching as their second choice. This position is being or has already been abandoned - at least by library users! It may be foreseen that in the future searching any library's information resources will happen through specialised search engines, for example Scholar Google and not through local library catalogues.

We are definitely not there yet, and local library catalogue searching should for a considerable time supplement engine searching. However, the trend is now clear, and libraries will not be able to stop it. Instead, they should focus on flowing with the trend intelligently and work practically to make their resources searchable and retrievable through search engines.

7. THE LEAN LIBRARY

The escalating costs of maintaining extensive, commercially based collections of electronic materials as well as a diminishing financial base will make it necessary for many libraries to make radical changes in their mode of operation. Self-service systems, maximum use of information technology, and rationalisation will be necessary. Also, libraries will have to focus on what should be their core functions in the future, and a number of traditional functions will be handed over to suppliers. The result will be a continuous reduction of staff resources used on internal processing and routine functions.

Library work will be a front-end job, in direct contact with users, with important consequences for the size and composition of staff. Academic library structures should be simple and effective and reflect the needs and structure of the university rather than parochial considerations of internal library functioning. Decision structures of the university libraries should be highly tuned to the needs and decision structures of the university.

8. CONCLUSION

Prophesying about the future is a difficult thing, and I do not pretend that the viewpoints, which I have presented today, are true - they are viewpoints! My main thesis is not that our library future will be this or that particular one. Rather, my main thesis is the following: important and to some extent irresistible forces are at work which will radically change academic library work in the future. If we ignore these drivers of development, we shall eventually disappear. If we understand them, we can intelligently choose the directions which will ensure our relevance and usefulness to academic research, teaching and study also in the next information age which will be the period of the internet, of the search engines and - eventually - of the disappearance of print media as the major medium of scholarly information.

Chapter 4

The Development of the China Digital Library

Zhang Tong

Yang Guihua, Associate Research Librarian Tianjin Library, The People's Republic of China

ABSTRACT

This article describes the China Digital Library (CDL) and explains how to use the computer network in China. It also explores the advantages of the China library in establishing the CDL. Finally, it examines the future construction and development plans for the CDL.

China is a large country that contains 5% of the world's population. Along with the rapid development of the Internet worldwide, China has advanced her steps. The digital library has become a focal point in the high-tech world. At the same time, it is also a very important symbol with which to evaluate the basic facilities of information sciences in a country. The leaders in library fields have done a great deal to play important roles in realizing the China Digital Library (CDL).

1. THE CURRENT SITUATION OF THE CHINA'S COMPUTER NETWORK (CDL)

China's computer network usage has become more and more common in recent years. Five national networks have been opened (see Table 1).

Table 1.

Network	Management Unit	Completed Time	Nature
CSTNET	China Academy of Science	1994, 4	Science and Technology
CHINANET	Information Industrial Ministry	1995, 5	Business
CERNET	Education Ministry	1995,11	Education
GBNET	Information Industrial Ministry	1996,9	Business
UNINET	Information Industrial Ministry	1993, 3	Business

Today, 200 universities and colleges in 8 regions in China are connected to these networks. According to calculations given by Chinese authorities, at the end of June, 2000, the number of computers connected to Internet was about 6.5 million, and the number of users accessing the internet was about 16.9 million. Based on these estimates, the number of users accessing the Internet will reach 30 million[1] in 2002.

2. THE APPLICATION OF COMPUTER NETWORK IN CHINA LIBRARIES

The Document Digital Center of the China National Library (CNL) was established in March 1999. At the China National Library, which is the fifth largest public library in the world, the resources have been digitized in CD-Rom storage. By the end of 1999, a total of 30 million pages, or 150 thousand copies of books, could be read on the Internet. In addition, 600GB of databases, catalogues, titles, and full text are available on the Internet at the CNL.[2]

The digital library set up jointly by the Liaoning Provincial Public Library and IBM supplies a large number of multi-medium databases with 10GB of full texts, pictures, films and TV. The contents include rare books, scenic images, cultural information, local documents, etc. Today, the Shanghai Public Library is creating a CD-Rom of rare books. 120 thousand copies out of total 170 thousand copies of rare books[3] will be planned to make CD-Rom. In the near future, rare books will be read by computer screen. Since the 1980s, The Integrated Library Automatic System (ILAS), designed and run by the Cultural Ministry, has unit users of 280.[4]

Many kinds of techniques are used to digitize the Chinese resources, for example: 1. Making a CD-Rom database by the traditional method of typing by hand, as the CD-Rom database of *People's Daily 50 Year*, etc. 2. Cooperating with some companies abroad. The Liaoning Provincial Public Library cooperated with IBM to protect the rare books with the modest techniques of encryption and digital photography. 3. Using scanners to input digital pictures. So far, many libraries in China have begun to order CD-Rom at home and abroad. It is very common to use CD-Rom databases in public libraries. For example, the Tianjin Library, being a city public library, collects many kinds of CD-Rom databases, such as: Academic Search Full text International published by EBSCO; Full text of Academic Journals published by Qinghua University, which contains more than 5,000 kinds of subject journals; Copy Materials of Full Text By Renmin University, which collects all kinds of key social science journals; and the Complete Library in Four Divisions, a CD-Rom database that contains one full Chinese rare book.

Encryption and scanning techniques have not been widely used in China. Therefore, the process of digitizing resources is very slow. We have a long way to go to compare with foreign countries. We must pay more attention to studying and developing the digital technique to meet the needs of users. The Share System Project on Document and Information Network, run by the Document and Information Center of the China Academy of Science, began in 1996 and was completed in 1998. Sharing documents and information from the Local Area Network, Metropolitan Area Network, and Wide Area Network has been realized. Since then, the network and resource construction of documents and information have been combined together and realized as the Hundred Units Network Project of China Academy of Science. Under such conditions, four regions and more than 100 institutes can be serviced with a catalog and information search, and remote access users can get requirements free of charge through CHINAPAC(China Packet-Switched Data Network) or the Public Telephone Network.

In Oct. 1997, The China National Public Library realized 2M connections with the China Public Computer Network and access to the Internet. In February 1999, the Library Area Network with 1000MB was put into use. The jointed point of Library Area Network is up to 2913. A 1000 M channel was put into use in March 2001 which is special for the China National Public Library. According to the calculation, more than ten projects on various digital libraries are being studied, such as the China Experimental Digital Library, the Intellectual Network-Digital Library System Project, mass storage and digital library application, etc. We can use these networks in various departments within China's public libraries, for example, acquisition, catalog and classification, circulation, information search, inter-library loan, international libraries exchange activities, etc.

3. PREPARATION FOR ESTABLISHING THE CDL

To catch up with the development of digital libraries worldwide, research work and development started in the mid-1990s in China. The concept of the digital library was introduced at the IFLA (International

Federation of Library Association) Conference in 1996, which was held in Peking, China. At the conference, IBM and Qinghua University demonstrated the IBM Digital Library Plan. In July 1997, the China Experiment Digital Library Project was reported to the China Planning Committee. Six public libraries participated in the project, including the China National Library and the Shanghai Library, among others. With this project, the CDL construction was officially underway. The Digital Library Key Project-CDL Demonstration Project is a large cultural project in which many organizations in China participated. The project was begun in 1999.

In October 1998, the China National Library Project was started by China's Cultural Ministry and the China National Library, In March 1999, the National Library Document Digital Center was established, with over 30 million pages of resources scanned annually.[5] In June of the same year, a CDL International Seminar was held in Beijing, China. At the same time, the network condition of CDL research work has reached the advanced level and technique has caught up with the international advanced Next Generation Internet (NGI) level. In this case, many research projects of the digital library began in China public libraries at provincial and municipal levels. The working meeting on the Library Project Resources Construction of China Digital Year was held at the end of 2000.[6] The First Plan for the China CDL Project (2000-2005) was discussed and worked out at the meeting and also determined to process the resources according to the international standards concerned. Digital library research institutes have been established in some universities such as the Beijing University and the Normal University in the Northwest China.[7] At the beginning of the 21st century, a new upsurge of research and construction of the digital library has started. From this we can see that the digital library research and construction work has developed so fast both in public and university libraries in China.

4. THE ADVANTAGES OF THE CHINA'S PUBLIC LIBRARIES IN ESTABLISHING THE CDL.

First of all, a library contains rich documents and information, which are important national resources. The levels of storage, collection, development and use of document and information resources of a country is the important symbol showing a country's cultural history, intellectual storage ability, and science and technology capacity. As one of China's largest social information systems, the information resources of a library have an important position that cannot be replaced in quantity or quality. After many years of development, several large library systems have been established in China: public libraries, school and university libraries, and special libraries. Each has its own characteristics in information construction but can work together. The China National Public Library has been engaged in establishing the largest Chinese information base. So far, several Chinese multi-media information resource databases, like "Poems," "China's Ancient architecture," "Marine Life," "Universe Structure," etc., have been completed.

Cross retrieval can be done through five successive resource databases. That is, you can access related databases from required resource databases very easily and conveniently to get the information needed. Besides this, some local public digital libraries have been established, such as the Shanghai Digital Library, in which the construction project contains the Shanghai picture dictionary, Shanghai cultural dictionary, rare books, Chinese newspapers and magazines, foreign language journal catalogs, etc.

Started in July 1999, the China Experimental Digital Library Project was completed in December 1999. The libraries participated the project have since set up their own digital libraries: Beijing University digital Library, Qinghua University Digital Library, Liaoning Provincial Digital Library (public library), Shanghai Digital Library (public library), and Tianjin Digital Library (public library). Some of each library's resources have been digitized, and we can read their special collections online. The CDL network will give us a wide concept of China's library that mainly supplies information and resources in Chinese. It can provide the services of reading, retrieval, and researching documents of special subjects online to the public.

Now, the top design of CDL Project--China Cultural Information Network--has been started. As calculated by the system of public libraries, at the end of 1999, the total quantity of documents in public libraries is up to 390 million copies, including large amount of rare books, local documents, nationality documents, special collection documents worthy of valuable research, etc.[8] These documents are distributed all over China. They are the social information system with the greatest quantity and used widely, which can be served to the people directly. The problem is how to develop them and use them through a network. From this, we can see clearly that the library is the one of the most important organizations needed to realize network services.

5. THE GENERAL PURPOSE AND DEVELOPMENT PLAN OF THE CDL PROJECT

The Digital Library Project in China has just begun. It is in an embryonic form that displays the library's transformation from a traditional library to a digital library, which gives us a hybrid library today. We know very clearly that to establish a real CDL, we have a lot of work to do. CDL is a large project across areas, across departments, across professions. It is also a newly created high-tech project.

High quality Chinese Digital Resources Databases Groups have been developed and are servicing China as well as the world through China Key Communication Network access to the internet. The techniques online in China have already been close to the international advanced techniques.

The Project Target

More than ten databases of Chinese digital resources with a total capacity of not less than 20TB will be completed by 2005.[9] At the same time, we will purchase many special databases abroad and realize online acquisition and classification, as well as inter-library loan in most parts of China. The digital library application system with Chinese distinctive features will be completed. Many high level professional specialists will be trained to continue the development of the CDL Project.

The Resource Construction of the Project

The resources Construction of the project include many fields, such as digital resource construction, system development, standards and specifications, and personnel training. We will develop a large number of Chinese multi-media digital resources databases with a total capacity of over 20 TB in the fields of culture, science and technology, and education.[10]

The Project Feature

The feature of the project is the virtual network, which is a large-scale, open, and disseminated digital information resources network system structure. It has many functions, for example: complicated information process and access; mass capacity information development and memory; Besides Chinese language, there are many other languages online including information resources on foreign languages and nationality languages; the network system with compatibility, good inter-operation, open distribution, etc. All the work we have done must meet the international standards and specifications. CDL research in China started late as compared with developed countries. To realize the real digital library, there are many inhibiting factors, both technical and non-technical. The technical factors that restrict the realization of the digital library include: 1. A very large amount of information to be digitized. 2. Without the unified agreements, specifications and standards. 3. The construction of a network infrastructure. 4. Safety problem of a digital library and copyright questions. The non-technical factors are questions of economy, management, and personnel. We need some time to solve all these problems, which are the same throughout the world. China, with its rich culture of 5000 years, is one of the largest countries in the world. We have the duty to do our best to share our information resources with the world and make the world we live in smaller and smaller.

REFERENCES

1. "The First Period Plan of CDL Project Construction" (2000-2005)
2. Deng Shengli "The Investigation of the Development of the CDL" (Information and Document Work 2001,3)
3. Liu Xiaoying "The Preparation at the Beginning of Digital Library" (Library Scholar Journal 1998,3)
4. Ibid (Note 3)
5. Yang Peichao "On Development Policy of the CDL" (Information and Document Work 2001,3)
6. Ibid (Note 5)
7. Ibid (Note 5)
8. Ai Qingchun "The Speech on the Meeting of the CDL Project Construction" (2000,12,17)
9. "The First Period Plan of CDL Project Construction" (2000-2005)
10. Ibid (Note 9)

Chapter 5

Academic Social Networks and Open Access: French Researchers at the Crossroads

Christine Okret-Manville

PSL Research University, Université Paris-Dauphine

ABSTRACT

Researchers benefit from an increasing array of tools to enhance direct communication and the dissemination of their research findings. These include Open Access repositories, Open Access journals, or hybrid publishing. For some years, researchers have been using new ways to communicate and share their work by using academic social networks.

In an attempt to foster the development of Open Access in France, the French consortium COUPERIN (Unified Consortium of Higher Education and Research Organizations for Access to Numerical Publications) proposed that academic social networks could be used to convince researchers of becoming more involved in Open Access. To test this hypothesis, a nationwide survey was launched in 2014 to explore whether and how these academic social networks are used to share content, but also how they compare to other Open Access classic tools. Within a month (20 May to 20 June), 1,898 researchers answered this 28-question survey. It was fully completed by 1,698 of them. This provides COUPERIN with considerable data for analysis. The respondents roughly reflect the composition of the French academic community in terms of gender and research fields, with a slight overrepresentation of young researchers/ PhD candidates.

This survey does not, however, cover the in-depth opinions of researchers on Open Access and academic social networks. It therefore only presents general tendencies. Nonetheless, the survey gives many indications as to how researchers apply Open Access. In addition, it shows how they feel about the usefulness of these networks compared to repositories when efficiently disseminating their work. This survey also takes the differences between disciplines into account and characterizes behaviour and opinions according to the different disciplinary communities and their research practices.

Finally, this survey allows us to define the main characteristics of a tool which could meet French researchers' needs for scientific communication. The components of such an ideal tool dedicated to Open Science could include efficient repositories to easily disseminate work and improve visibility, a sharing network and the scientific stamp of peer-review.

KEYWORDS

Open Access; Academic Social Networks; Researchers; France; Perception; Exploratory Survey; Open Science

1. INTRODUCTION

For many years, the development of scientific communication has been a focal point for researchers, inducing the development of Open Access. As part of this movement specific tools have been created aimed at easily sharing research findings. These have put researchers in touch with one another, have bypassed the sometimes lengthy delays between the initial submission of a paper and its publication, have developed direct peer review, projects, and have improved collaboration within the academic community. An array of possibilities exist online for researchers to disseminate their work such as Open Access repositories, Open Access journals or hybrid publishing.

In France, publishing in Open Access is not (yet) a public requirement for research. Academic assessment neither takes Open Access publishing into account. The Minister for Higher Education and Research Mrs Geneviève Fioraso expressed an official support during the Open Access Days of January 2013.1Although she presented a roadmap for Open Access, advocating the benefits of publishing and sharing Open Access to researchers mainly takes place in research and higher education institutions.

Consequently, the Couperin consortium – composed of all French research organizations and universities – has been active for some time in this field. It contains working groups on IPR, technical matters, advocacy (a website dedicated to Open Access for researchers to open in October 2015) and best practices. It also organizes Open Access Days every two years, which aim to evaluate the progress of Open Access in France. They serve as a platform for debate. Looking for the best way to reach researchers and tell them about Open Access, Couperin decided to organize a survey to better understand their work practices and opinions related to Open Access (**Joly, Okret-Manville, & Vignier, 2014**). In addition, recent years have seen a significant development in social networks with some dedicated to researchers. Couperin thought that it would be useful to know how these networks were being used by researchers and to evaluate whether an efficient Open Access advocacy policy could be pushed through this channel.

2. METHOD

To prepare this survey, 3 sources were used: reports dealing with the specific link between research and social networks, discussions on blogs and forums, and discussions with librarians. This content helped define the mood of the academic community about these tools and provided a list of questions, positive and negative remarks that could be used for the questionnaire. At the time of preparation (1st quarter of 2014) not many studies existed on this new subject. The major part of this documentation is in French.

This resulted in a set of 28 questions, which was divided into 3 parts. The first set of questions focused on the usage and practice of social networks (general social networks and then more specifically academic ones), the second one dealt with the perception of Open Access. The third part aimed at pinpointing the academic profile of the respondent (gender, age, research domain, affiliation). Free comments were also welcome.

This questionnaire was available online from the 20th of May to the 20th of June 2014. The link was sent to the Open Access contacts Couperin has in every university and research organization library. Libraries in turn disseminated the questionnaire to their researchers and actively promoted it.

1898 responses were obtained. Among them, 200 answers were incomplete, ending up with 1698 results to exploit, and 700 comments. Given the length of the questionnaire, this can be considered a fairly good outcome.

Our sample can be deemed reliable, as respondents are rather representative of the French academic community. As regards gender, we notice a slight overrepresentation of female respondents (42.9%, compared with 35% in academia). Two-thirds of our respondents are under 39 years old. This gives an overrepresentation of young researchers, PhD students included, as in France researchers over 39 account for 55% of the research community corpus. A breakdown by disciplines shows the overall correspondence with the French academic landscape, with a slightly stronger representation of life sciences (Table 1).

Table 1. Proportion of respondents compared with the population of French researchers broken down by macro disciplines.

Macro disciplines	Respondents	French researchers (2013)
Economics/Business/Law	12%	13.5%
Social Sciences (incl. Arts & Humanities)	25%	27%
Hard Sciences (incl. Mathematics/Computer Sciences, Physical Sciences, Engineering Sciences)	44%	45%
Life Sciences	19%	14.5%

A large Excel sheet was produced for this analysis where all answers were entered and sorted, particularly to identify possible disciplinary effects. Respondents had to choose their research discipline from 30 possibilities. We chose to cluster these disciplines into larger domains for more efficient analysis. Seven disciplines were consequently used: Economics/Business/Law, Arts and Humanities, Social Sciences, Mathematics/Computer Sciences, Physical Sciences, Engineering Sciences, Life Sciences. We also assume that the clustered disciplines are similar or very close in nature related to the theme of the survey.

3. RESULTS AND DISCUSSION

3.1. Researchers and Academic Social Networks: Potentials and Questions

The majority (60%) of our respondents have heard about academic social networks although only 42% use them. They usually choose ResearchGate (by far the most popular (65%)), and then Academia.edu (24%). Other academic tools like Mendeley were mentioned by 2.1% and 1.9% for Google Scholar, with a long tail of other networks following behind. Since no predefined list was used, these answers suggest that some researchers cannot always precisely identify what an academic social network is. They may see them as *generic scientific communication tools*, including specialized search engines such as Google Scholar, reference managers such as Mendeley or open access repositories such as ArXiV. For this study, we consider academic online social networks dedicated to researchers where resources and information are shared within the academic community.

The 58% who do not use academic social networks are primarily not convinced of their usefulness ('unuseful' for 41%), don't have time to use them or feel they lack information about them (14%). Some find academic social networks have a too limited offer of services and 4% fear they would lose control over their professional and personal data by engaging with them.

A breakdown by discipline reveals a greater usage of academic social networks in the Social Sciences (48%) and Life Sciences (47%). When researchers were asked about their usage of social networks in general, they do not tend to use social networking without considering their domain of study first. Broadly speaking, Human and Social Sciences have adopted social networks to develop direct communication within their communities: researchers in Social Sciences, Arts & Humanities, Economics/Business/Law report a large use of Facebook, Twitter and LinkedIn. Social Science researchers are the most engaged in social networking here, be it for academic or general reasons. Otherwise, it is

Economics/Business/Law and the Engineering Sciences that favour general networks. At the other end of the spectrum, Physical and Natural Sciences use these tools less, having a more balanced usage between the different types of networks (Table 2).

Academic social networks are mainly used in community contexts. A third of the respondents mentioned using them to interact with other communities, shop for new ideas and to stay up-to-date with other subjects other than their core themes of research. This practice seems quite widespread in the Life Sciences (59%) and in the Arts & Humanities (46%) whereas in Mathematics/Computer Sciences (19%) academic social networks is first and foremost an 'internal' channel for research exchanges. The main use researchers assign to networking is to disseminate their publications (70%), especially in the Life Sciences and Social Sciences. We can also observe that sharing data is not negligible, particularly in the Arts & Humanities reporting more than 50%. It is somewhat paradoxical that researchers in this discipline share data via these networks as only 18% admit to knowing about the site's data-sharing policies. This can be interpreted as voicing a concern about the (supposed or real) lack of a local (or national) research infrastructure in these disciplines. This drives researchers to use existing channels to share data, regardless of any future changes to the conditions of access to their data.

Two-thirds of academic social network users are quite satisfied with them, and the degree of satisfaction depends on the discipline. The biggest users, i.e. in the Social and Life Sciences (see Table 2 above) are the most happy with academic social networks. The reverse is also true: Mathematics/Computer Sciences and Engineering Sciences report are most dissatisfied. Two assumptions can be made to account for this fact: 1) bad knowledge of these networks or 2) a lack of acceptance and practice with them in their communities. The second part of this study, which deals with the perception and use of Open Access, may give us further explanations.

Table 2. By discipline, users of general social networks and users of academic social networks.

	Arts & Humanities	Economics/ Management/ Law	Maths/ Computer Sciences	Physical Sciences	Life Sciences	Engineering Sciences	Social Sciences	All disciplines
Users of general								
Social networks	73%	82%	68%	67%	66%	73%	74%	71%
Facebook	64%	62%	47%	54%	52%	58%	64%	56%
LinkedIn	35%	62%	44%	41%	42%	50%	41%	45%
Twitter	21%	24%	22%	14%	13%	9%	25%	18%
Users of academic social networks	33%	33%	41%	40%	47%	37%	48%	42%

Respondents to the survey then indicated what the advantages of academic social networks were by choosing assumptions from a closed list (multiple answers possible). Researchers easily agree on the way academic social tools can serve their goals. As Figure 1 shows, the two main advantages they see in being active in such a network are to 1) share content and 2) to raise their scientific profile and academic reputation (60 to 70% respectively). Then comes collaboration with colleagues: getting in touch with them, doing collaborative research, creating project communities, getting new ideas (15 to 30% respectively). **Jordan (2014)** has noticed in her study on academics and their online networks that the primary objective of researchers who create a profile in these networks is to bring more visibility to their publications, and to use them as a kind of 'online business card'.

Figure 1. Advantages of academic social networks.

Limitations mentioned (see Figure 2) suggest a very critical eye, even when researchers are rather satisfied with the networks. The main criticism is a scattering of resources (35% of all responses). Around 25%

of answers were critical towards the absence of author protection, no peer reviewing, no reliability, and too few functionalities (i.e. related to data-sharing). Roughly 20% question the relevance of information posted and claim that access to information is difficult. Respondents' comments confirm these observations.

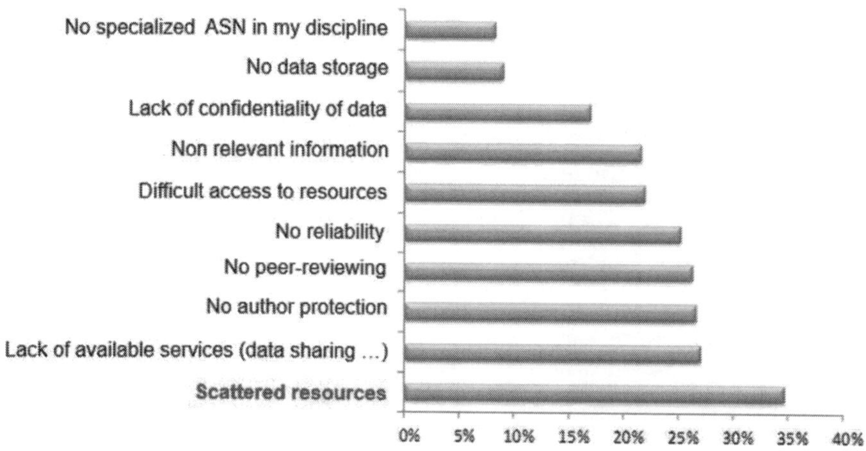

Figure 2. Limits to academic social networks.

These opinions reflect a good perception of the social networking side of these tools, but some deficiencies on the side of the academic workflow. To sum up, these networks appear as one among many different channels to make peer interaction easier, being notably used to integrate young researchers (e.g. PhD candidates) in research circles. They are first and foremost expected to boost the visibility of research works on the Internet. They are then appreciated as collaborative tools. These results qualify the views an analyst of an American media consulting firm expressed about these networks: 'These are tools that people are using to raise their profiles and become more discoverable, not community tools of social interaction.' (**Van Noorden, 2014**, p. 127) On a darker note, they have general terms of use where the privatization of data is a possibility, and for some they remain complicated to use and insufficiently secure for data storage.

3.2. Researchers and Open Access: Open Access Still has a Long Way to Go

Knowledge of Open Access is reported by 70% of all respondents; it is considered as an important feature of the research landscape today. However, differences exist between disciplines here. Across the spectrum, Mathematics/Computer Science and Life Science researchers have heard of it at a rate of 76% whereas Economics/Business and Law, only 54%.

Yet, depositing publications in Open Access repositories is not commonplace: 29% of all respondents report depositing, 41% do not deposit, and 30% do not answer this question.

The survey provides an overview of the French Open Access repositories landscape. The national repository HAL (Hyper-Article on Line), managed by the National Centre for Scientific Research (CNRS) and with strong political support, is mentioned by 71% of the answers. These answers encompass HAL itself and the sub-portals it has created at the request of many research organisations and universities. ArXiv is the second most important repository cited (19%), although it is impossible here to determine whether ArXiV is directly used by researchers or rather more indirectly using HAL to push data to it. 10% of all responses mention institutional repositories (DSpace, E-Prints, Flora, home-grown software ...) and some subject-based archives such as RePEc. This picture reflects the dominance of HAL, as a specific French answer to enhance Open Access.

Open Access deposit also relates to what is the norm in scientific communities. It is Mathematics and Computer Science communities who particularly engage in depositing their outputs in Open Access as do the Physical Sciences (Figure 3).**Creaser et al. (2010)** make the same observation and specify that for these disciplines, speed of dissemination is considered a priority. They add that mandates are not seen as important (p. 156), which suggests that depositing may reflect an adhesion to the principles of Open Access. Regarding Mathematics/Computer Science, a survey conducted in 2006 among French researchers of that domain reached the conclusion that this conviction was widely shared in this community.2

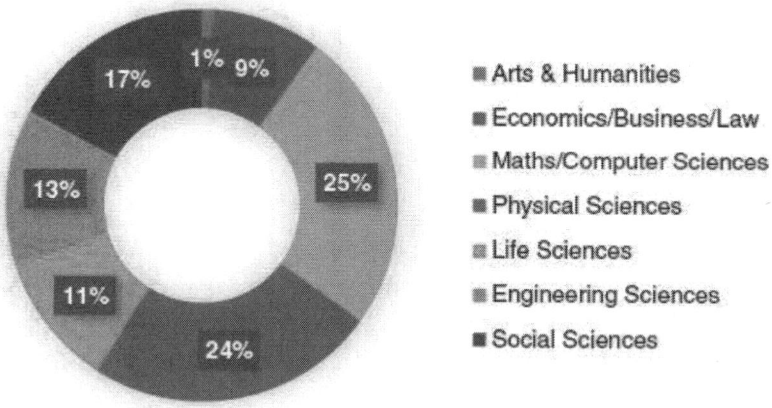

Figure 3: Based on respondents who deposit in repositories by discipline.

This graph therefore confirms our assumption about the profile of Mathematics/Computer Science scientific communication. It overwhelmingly favours 'traditional' Open Access repositories to disseminate their work and gain visibility. This seems satisfactory enough, so this community does not feel that they have to turn to academic social networks to help them reach this objective more efficiently.

In the Life Sciences, the rate of deposits is quite low (11%), compared with the high rate of respondents in this discipline who know about Open Access (76%). This is despite the fact that pushing publications to PubMed is easy via HAL. There are two explanations for this. Many medical journals have agreed to push publications to PubMed without researchers' intervention (**Creaser et al., 2010**, p. 155). Similarly, an agreement has been reached between some research institutions and the open access publisher BioMed Central, who automatically sends researcher metadata to HAL where the institution has a HAL sub-portal (**Prime-Claverie & Mahé, 2013**, 5th part). These automatic workflows make deposits effortless for Life Sciences researchers. A more essential explanation lies in the importance of peer-review in the Life Sciences, which is considered a crucial part of the academic workflow. This means that as far as dissemination of research outputs is concerned, Green Open Access does not guarantee the scientific value of documents. In the Open Access world, this can only be assured by Gold publishing (**Creaser et al., 2010**, p. 156; **Prime-Claverie & Mahé, 2013**, 5th part).

Researchers in Economics/Business/Law do not deposit much in Open Access; this may be linked to their reported poor knowledge of this movement.

As a general result, only 29% of respondents state that they deposit their publications in Open Access. The 71% who do not engage with this movement provided comments to explain their attitudes. These comments have been clustered by theme:

32% referred to a lack of knowledge on Open Access or to its ambiguity for their discipline. Comments often suggest the idea that Open Access 'costs too much'. This may be the sign of a confusion between Green Open Access and Hybrid Open Access where some researchers believe that taking up Open Access publishing is somewhat inescapable. The ambiguity of the notion of Open Access is seen in the Life Sciences (see above). Other comments, however, do indeed help identify a real lack of information about Open Access, its nature, the deposit process, existing repositories, their relationship to social networks or personal webpages, their advantages in terms of visibility and effects on publication accessibility.

A second set of arguments (19%) relates to lack of time. Depositing documents requires an extra-time these researchers are not readily willing to take because it diminishes the time they can dedicate to pure research. This suggests that respondents are satisfied with the traditional ways of promoting their works (through commercial publishing or by giving lectures), seeing depositing as some kind of red tape. The link between open access and citations increase is not obvious to them.

Some researchers (16%) also feel unsure about the legal possibilities of exploiting their work once they have submitted to a journal. They lack information about IPR, and the Sherpa/RoMEO site.

Not depositing a document can also be part of various publication strategies (14%). There is no official incentive to publish in Open Access unless this would be taken into account in the assessment of their career. Although Open Access journals may offer good quality content, they seldom have a high impact factor. In addition, researchers sometimes get pressure from publishers or from senior researchers (particularly PhD students) to favour commercial publishing.

Publication habits can account for a lack of involvement in Open Access as well (12%). This type of answer shows that these researchers do not ponder about the scientific communication workflow and neither do they question their habits as the whole process of scientific communication is changing around them.

Finally, some perceive the complexity of the deposit process of some repositories as an obstacle for joining the Open Access movement (7%). Those who express this view also oppose 'easy to deposit' sites such as academic social networks.

Many comments also point to the fact that there is no clear political support for Open Access on a national level. These comments show significant support for Open Access. They suggest three levels where a national policy could make headway: strongly encouraging (or mandating) researchers to deposit outputs of publicly-funded research in Open Access, taking into account Open Access journal publishing in career assessments, putting pressure on publishers to promote self-archiving. Comments also argue that universities and research organisations should be mindful to actively foster Open Access in their structure.

As regards other means to freely disseminate research, the survey also shows that the Golden road has not yet made an inroad in the researcher's publishing practice. 71% of respondents did not answer the question relating to this theme. This may mean that they are not familiar with Open Access journals in their discipline and may not be very interested in them as assessment policies *de facto*foster publishing in commercial journals.

All in all, 11% of respondents reported having already deposited in Open Access journals. A breakdown by discipline should be analysed with caution due to the low response rate. However, it can be said that depositing in Open Access journals seems to make particular sense in the Life and Physical Sciences. This concurs with these disciplinary practices (see above) where some know about APCs and pay them.

Mentioning Open Access journals brought up many critical remarks about publishers' position in the academic workflow with, again, a confusion between Gold and Hybrid. Some comments also expressed worries about the scientific quality of these journals.

Beyond sharing publications, sharing data is becoming a relevant issue for researchers. Asked about the storage of their data, 83% of respondents keep them on their hard drive, and 40% use this as their sole storage solution. 22% mention data sharing sites (few cited sites:

ArXiV, HAL, Github ...); It was the Social Sciences and Arts & Humanities who mentioned sharing their data most (at around 20%).

3.3. Academic Social Networks Versus Open Access? A Comparison of Perceptions

In the third part of this study we compare the perceptions of academic social networks and open access regarding four criteria: visibility, publication dissemination, data protection and data preservation. Figure 4 illustrates general results.

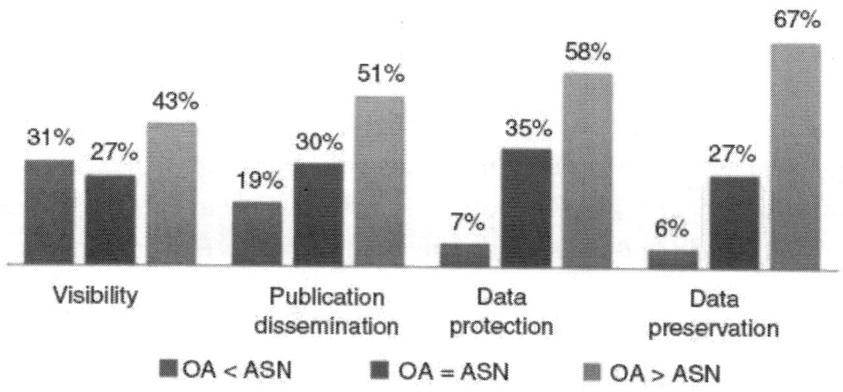

Figure 4. General comparison of perceptions of academic social networks and open access repositories.

For all areas, Open Access sees the more positive responses. Open Access repositories seem to fulfil a legacy role, keeping access to data secure and preserved for the future. This long-term position is their main strength. For publication dissemination, Open Access is at an advantage. But as far as visibility is concerned, Open Access is challenged by academic social networks in researchers' minds.

Breaking down these results by discipline, academic social networks already take the lead role on this issue in Life Sciences and Engineering Sciences, and are even with Open Access repositories in the Social

Sciences (Figure 5). Apart from this, Open Access remains seen as more efficient than academic social networks.

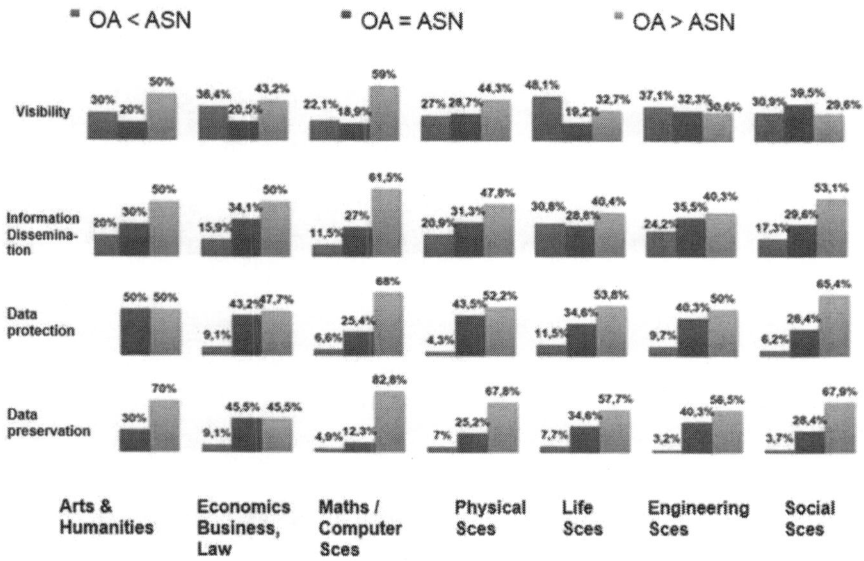

Figure 5. Comparison of perceptions of academic social networks and Open Access repositories by discipline.

The degree of conviction varies according to discipline: in Mathematics/Computer Sciences Open Access unquestionably outperforms networks, but in other disciplines, this perception is more mixed or differs in relation to the different criteria. In the Arts & Humanities and Economics/Business/Law, confidence in academic social networks for data protection and data preservation remains surprising given the lack of knowledge researchers in these domains have reported here about the policy of data usage for these networks.

Using the data of this study, we can divide researchers' perceptions into two main groups. The first one includes the disciplines that rely on Open Access processes to disseminate and freely share scientific output making complementary use of academic social networks. Mathematics and Computer Sciences concur exactly with this description, as do Physical Sciences though with a little less enthusiasm. Engineering Sciences are a step behind: even though they use Open Access, they use academic social networks to boost the visibility of their outputs. Life Sciences stay true to the practice of peer-review. This makes them pay more attention to the Golden road although they use networks for dissemination and to shop for new ideas.

The second group tends to make a larger use of academic social networks, i.e. for the Social Sciences, or in the Arts & Humanities. These groups use them to get updated about different subjects or to share data.

Economics/Business/Law find themselves in between these groups.

Accounting for this difference in perception is difficult today and would require complementary in-depth studies. However, as far as raising awareness about scientific outputs is concerned, it looks as though there have been two waves of involvement in Open Access. The earliest wave includes researchers from the Natural and Physical Sciences who have adopted Open Access repositories and it seems that these continue suiting globally their need for direct scientific communication. The development of new dedicated social tools does not seem, at least today, to seriously question this workflow. The latter group, generally made up of the Humanities and Social Sciences, report a clear interest in new ways of disseminating their work using academic social networks. They question the ability that Open Access repositories have to give their output the maximum visibility.

In answering this survey, researchers have left a hundred comments which allow us to conclude on what their dream tool for Open Science could be. These very comments show that for many researchers neither repositories nor academic social networks are completely satisfactory ways for promoting or sharing research. The ideal tool should be: national but also open to international usage, not used for assessment, using open technologies, user-friendly, covering the whole publishing process including peer-review, providing access to publications and data, offering various services (calendars of events, calls for papers, secure storage for research, forums, videoconferencing tools, etc.).

4. CONCLUSIONS

For the Couperin consortium, this survey has established a set of issues that librarians have to work on to locally increase awareness about Open Access among researchers. It is therefore important to push information about Open Access and particularly issues such as publishers' policies and IPR. Academic social networks could help reach those who favour the use of these tools. We must then support researchers by tailoring services designed to help them increase their visibility without infringing too much on their research time, by mediating deposit in multiple relevant repositories. For example, young researchers and PhD students must be targeted first to give them the keys to an easy navigation in a radically changing academic world and to draw a picture of what awaits. Even though we did not single out PhD students among the researchers, many of them answered this survey. It has been observed that junior researchers (under 39) deposit in Open Access and use academic social networks less than senior researchers.

Supporting Open Access in France would certainly be more efficient if action was taken on a national level. Researchers have expressed their views about this. A law on digital issues is currently being prepared by the Government, and it is hoped that it will include a clear policy in favour of the open and free dissemination of French research output.

The second lesson learnt from this survey is the clear complementarity between Open Access and academic social networks. The dream tool for Open Science described by researchers confirms this. Testimonies and thoughts shared on blogs or in conferences highlight the 'winning combination' of academic social media and research (**Shepherd, 2014**).

Academic social media echoes research of the academic world more powerfully. This gives much food for thought to elaborate local strategies to help increase researchers' visibility.

However, this is an exploratory study, which produces an overview of the perception of researchers about academic social networks and Open Access. It would require further studies to gain a more precise insight on disciplinary bias. Differences were identified between disciplinary communities despite clustering them to make our analysis easier.3 It could also be interesting to make this survey again in a few years from now to trace possible evolutions, as a result of the ever-changing landscape of communication and dissemination tools and perhaps as a mark of the effort made to promote the free dissemination of scientific French output on the Internet.

ACKNOWLEDGEMENT

The complete analysis of this survey was co-written with Stéphanie Vignier and Monique Joly (**Joly et al., 2014**). Many thanks to Vanessa Proudman and André Lohisse for their suggestions and comments on this paper. The views expressed in this article do not reflect any official Couperin position. Conclusions are mine.

REFERENCES

1. Creaser, C., Fry, J., Greenwood, H., Oppenheim, C., Probets, S., Spezi, V., & White, S. (2010). Authors' awareness and attitudes toward Open Access repositories.*New Review of Academic Librarianship*, *16*(S1), 145–161.

2. Graziotin, D. (2014). Green open access in computer science – an exploratory study on author-based self-archiving awareness,

practice, and inhibitors.*ScienceOpen Research,* June 16, 2014. Retrieved August 4, 2015

3. Joly, M., Okret-Manville, C., & Vignier, S. (2014). *Réseaux sociaux de la recherche et Open Access. Perception des chercheurs. Etude exploratoire.* Retrieved August 4, 2015, from http://www. couperin. org/groupes-de-travail-et-projets-deap/open-access/286-open-access/1214-reseaux-sociaux-de-la-recherche-et-open-access.

4. Jordan, K. (2014). Academics and their online networks: Exploring the role of academic social networking sites. *First Monday, 19*(11). Retrieved August 4, 2015

5. Prime-Claverie, C., & Mahé, A. (2013). *Sites de dépôt en libre accès et formes de médiations: quelles évolutions?* DocSoc 2013 – Diversification et renouvellement des médiations – Zagreb, apr. 2013, Croatia. Retrieved August 4, 2015,

6. Shepherd, E. (2014). *Research-Open Access-Social Media. A winning combination.* Open Access Week Symposium, Rhodes University, October 2014. Retrieved August, 4, 2015, from http:// www. slideshare.net/EileenShepherd/researchopen-accesssocial-media-a-winning-combinationhttp://fr.slideshare.net/ EileenShepherd/ researchopen-accesssocial-media-a-winning-combination

7. Van Noorden, R. (2014). Online collaboration: Scientists and the social network.*Nature, 512,* 126–129. doi: 10.1038/512126a.

8. Wojciechowska, A. (2006). Usage des archives ouvertes dans les domaines des mathématiques et de l'informatique. *Documentaliste – Sciences de l'information,43*(5–6), 294–302. doi: 10.3917/docsi. 435.0294.

Chapter 6

Getting to Know Library Users' Needs — Experimental Ways to User-centred Library Innovation

Karen Harbo, Thomas Vibjerg Hansen

ASB Library, Business and Social Sciences, Aarhus University

ABSTRACT

'Meeting the User' is a programme committee under the Danish Electronic Research Library. As a development group at a national level we see our role as facilitating an innovative culture within academic libraries, focusing on users' needs and the way libraries meet them.

In collaboration with a consultancy firm, the committee organized a travelling workshop in four cities in 2010. The workshop introduced practical ways for library staff to get to know their users' needs for services and was based on anthropological methods. The travelling workshop was part of a larger project called 'A Journey of Discovery in Danish Library User Land' (translated from Danish Brugerkaravanen), which also included a national thematic day, a blog and the publication of a method guide.

There is an ongoing need for academic libraries to improve their services. One strategy is to become more aware of the users' needs. On the one hand we have libraries which give access to a lot of information, offer courses in information literacy and strive to be a part of the learning environment. On the other hand we are not always certain of the needs of our users.

If libraries want to improve the way they serve their users' needs, they must innovate their services, facilities and courses by building upon what you could call 'user logic' and not upon classical 'library logic'. 'User logic' is that which is meaningful for the user instead of what is traditionally meaningful for a library. Finding out what the users' needs are, requires methods to study the users. In order to discover the shortest route from knowledge via idea to action, we looked for methods that can be employed by librarians or library information specialists.

This article describes how 110 librarians and information specialists acquired such methods in four cities and in four days, by means of a workshop structured like a guided tour through the land of library users. The goal of the article is to give other libraries inspiration for ideas, concepts and concrete tools to study user behaviour and become more aware of the user's needs for service.

The article contains a description of the above-mentioned methods, valuable experiences from the workshop, a presentation of the concept and concrete tools, discussion of the concept of user logic and library services, and the seven principles for human-centred innovation in relation to libraries, a short list of studies carried out by librarians and discussion of further perspectives.

KEYWORDS

user-centered innovation; user needs; library innovation; anthropological methods; user logic; Brugerkaravanen

1. INTRODUCTION TO 'MEETING THE USER'

The Danish Electronic Research Library (DEFF) is an organization which is co-funded by three Danish ministries: the Ministry of Science, Technology and Innovation; the Ministry of Education; and the Ministry of Culture. At the national level, DEFF collaborates with the institutions under the three ministries that co-fund DEFF, primarily libraries, academic institutions and secondary educational institutions. Internationally, DEFF collaborates with partners both in the Nordic region and the European region. The DEFF steering committee is responsible for DEFF's strategy and actual activities.

Four groups monitor DEFF's large domain of interest, which includes initiating projects (based on subsequent steering committee approval) and coordinating projects across the groups and with DEFF licenses. One of these groups, 'Meeting the User', worked as a development group at the national level from 2008 until 2011. The group's three main objectives were:

1) To facilitate an innovative culture within academic libraries focusing on users' needs and the way libraries meet them.

2) To initiate projects in the fields of information and knowledge creation, learning and communication at the junction between the physical and the virtual library environments.

3) To encourage sustainability of the project outcomes.

2. INNOVATION IN LIBRARIES

There is an ongoing need for academic libraries to improve their services, if only because of the considerable amounts of resources spent on library services and the need to ensure value for money. Libraries must take strategic action in order ensure that the services offered are the services needed. One strategy is to become more aware of the users' needs for services. On the one hand we have libraries which give access to a lot of information, offer courses in information literacy and strive to be a part of the learning environment. On the other hand we are not always certain of the needs of our students, teachers and researchers. What is traditionally meaningful for the library is not necessarily meaningful for the library user. There is always a risk that what libraries regard as the right answers to user needs, may be answers to questions that were never raised users in the first place.

In addition, users' needs may change imperceptibly over time and libraries must constantly take such changes into account. For example, today's students may still need guidance in using the enormous amount of e-based literarture, but at the same time they are, beyond dispute, advanced users of the latest information technology which library institutions, often conservative by nature, are only beginning to experiment with.

Offering services that are of no need to users is an expensive business. One ambition of the DEFF programme committee 'Meeting the user' was to challenge the state-of-the-art mindset of libraries and library staff by encouraging them to switch from what you could call classical 'library logic' to 'user logic'. Another ambition was that the new insights and knowledge gained by project participants could be transferred easily into their own library settings.

3. THE CONCEPT OF USER LOGIC AND LIBRARY SERVICES

Does a particular service appear to be logical and natural to a user, or does it appear to be logical to the library? This is the simple test question we ask ourselves when we innovate our library services. The next question is whether it should be either, both or neither.

Libraries support researchers and students in their research and learning and it is our job to allow them to benefit from the library's access to materials and services. In other words it is about the library's front end services. The question arises whether these should be based on the library's perspective of proper use of resources or on the user's perspective? In our terms we talk about library logic vs. user logic. The latter is our focus area!

User logic is about meaning. Are our services meaningful to our users from the users' point of view or from that of our libraries? Do our services appear to be logic to the users? Do they match the way users work?

The concept of 'user logic' is related to professor emeritus Carol C. Kulthau's description of the user's perspective on information retrieval and the paradigm shift from source or bibliographic oriented services to process-oriented library service (Kuhlthau, 2004). Briefly, Kulthau distinguishes between searching for sources (source oriented paradigm or library logic) and seeking meaning (process oriented paradigm or user logic). Sometimes library logic and user logic match, sometimes they do not. We have to remember this when we innovate our libraries and library services. We must also remind ourselves that the context is crucial: the kind of educational institution, the types of students and their learning background, the types of researchers and their topics, the educational culture and traditions, the social culture, etc.

A user logic point of departure means that we place the user in the centre, not the library. We do not focus on introducing the library's collections, but we focus on helping the user along in his or her research or study process from a holistic point of view. The library's collections are part of the services we offer, but they are by no means the only services (Figures 1 and 2).

Figure 1. Library logic with the library in the centre.

The classical library catalogue based on a classification system (e.g., UDK, Dewey) is a typical symbol of library logic. It forces the students and researchers to use the library in a particular way. Different libraries have different classification systems, even though they have the same kind of students!

The concept of user logic contrasts with library logic. It is not a well-defined concept, rather a concept we use for critical awareness when we want to innovate library services. We ask ourselves whether we want to innovate services, support, access, teaching, counseling etc. according to library logic or user logic. It is difficult to define a clear-cut rule about what is user logic and what is library logic. However, some symbols of library logic could be those shown in Figure 3.

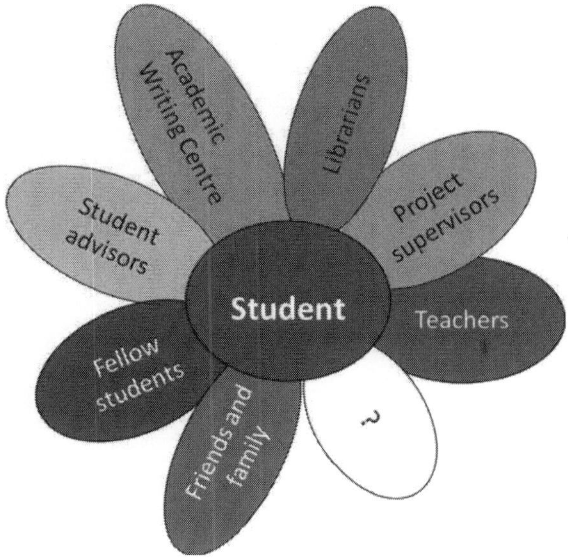

Figure 2. User logic with the user in the centre.

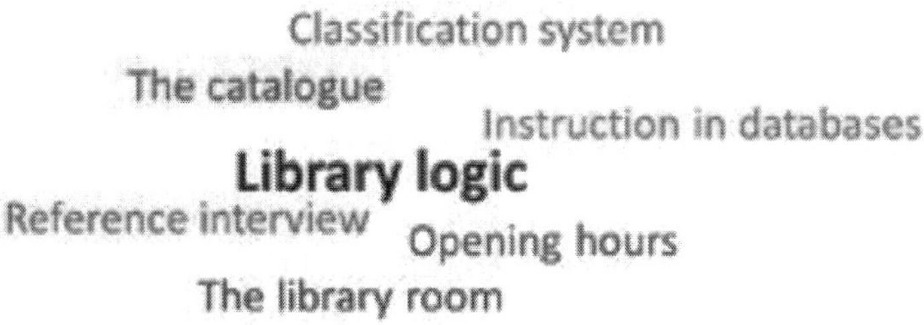

Figure. 3

Some examples of user logic test questions are:

1) Do you focus on the user's process or on the library's information resources in your library teaching, information counseling and other contacts with users?

2) Does your library reflect the users' needs, e.g., in terms of work spaces, facilities, access to materials, and resource or process-oriented help?

3) Do you understand how your users work or study?

4) Does the user require help to move along his studies or does the user need to know everything the library can offer?

5) Is the location of the service logical for the users? Should the services be available in the library where students and researchers work, or should they be available from the library web or another virtual place?

The methods, techniques and tools demonstrated in the project 'A Journey of Discovery in Danish Library User Land' help librarians to find answers to the above questions.

4. THE PROJECT 'A JOURNEY OF DISCOVERY IN DANISH LIBRARY USER LAND'

In 2010 the programme committee 'Meeting the User', in close cooperation with an Aarhus-based consultancy firm named 'Changepilot' (John Lassen and Rebekka Høy Biegel), decided to launch a large and experimental development project on a national scale: A Journey of Discovery in Danish Library User Land. The expedition was termed Brugerkaravanen (which means: a caravan setting out for library users). The Brugerkaravanen consisted of 110 library employees from academic libraries covering different levels of educational institutions. In sixteen mini-buses they visited 32 locations where users live their personal and professional lives in the cities of Copenhagen, Aarhus, Odense and Aalborg (Figure 4).

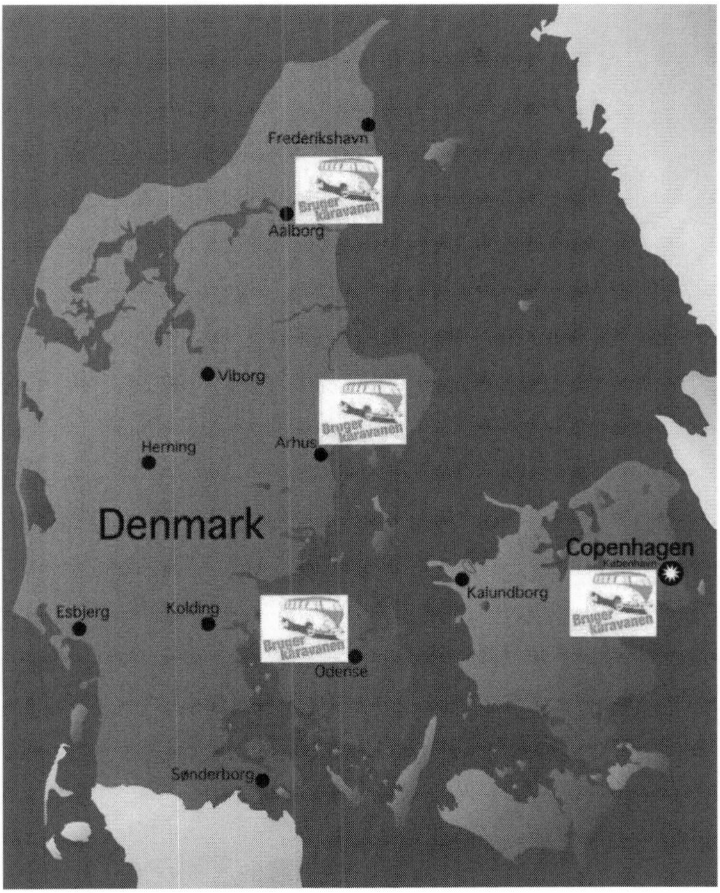

Figure. 4

In each of the cities meetings were arranged between the participants in the project and different categories of users. The project focused on four particular categories:

1) Master's students who already have completed higher education, are employed and have a demanding career.

2) Nursing students.

3) Exchange/international students.

4) Researchers in the field of arts and Humanities.

The consultancy firm 'Changepilot' that arranged the expedition to the four main cities of Denmark also arranged the meetings with these groups of users. In each city the participants were picked up in groups of eight by mini buses that took them to a 'base camp' for the tour of the day. By way of introduction the participants were shown a presentation about seven principles of human-centred innovation. After this they were introduced to qualitative methods and techniques inspired by the scientific fields of ethnography and anthropology: a set of user-centred tools for investigating users' needs. The presentations were intended as theoretical as well as practical background information to the programme of the day which included rides in minibuses to visit users that represented one of the above four categories.

The project had invited library staff to leave their (safe) libraries and go to where the users live, work, study, have coffee, listen to music, meet with friends and co-students. During these meetings with users in the field, the library staff were supported by expedition guides (employees from the consultancy firm). These helped the library staff apply the anthropological methods they had been introduced to, i.e., varied forms of interviews and observations. By interviewing and observing during the day, the library staff collected valuable data about the users. While collecting data they attempted to forget about libraries and classical 'library logic' and to focus on the people they met, not only as library users, but as ordinary people, living ordinary lives and doing ordinary work and or studies. By the end of the day, the data collected were analyzed, categorized and presented at the project base camps — still under the skillful guidance of the expedition guides.

After the four individual expeditions had taken place in the cities of Copenhagen, Aarhus, Odense and Aalborg, a final conference was held in Aarhus, to draw conclusions about the insights gained; these were presented as working questions by the participants by the end of each expedition. Many library staff participated in both events, but the final conference also attracted new participants. The conference took the work one step further and by the end of the day a publication was launched to document the methods and tools presented as well as a web site which also offered the participants a virtual space, a blog, to share knowledge on the subject in the future.

5. ONE PARTICIPANT OF THE AARHUS EXPEDITION SHARES HER EXPERIENCES

'Having signed up for the 2010 Journey of Discovery in Danish Library User Land in Aarhus, I had been told to meet in the hall of the central station at 9 o'clock. I had brought the little bright green suitcase that had been sent to me by the project, filled with items that had awoken my curiosity about the seminar on user-centred innovation which I was going to attend. While waiting I met other participants — easily recognized by the little bright green suitcases they all carried (Figure 5).

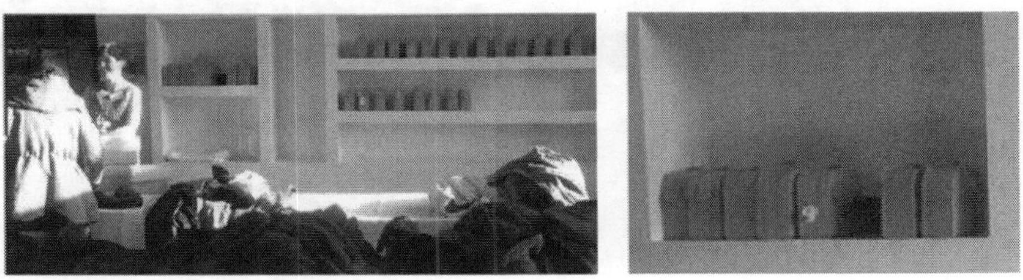

Figure. 5

'At 9 o'clock we were met by people of the project. A row of minibuses displaying bright signs with the name of the project waited for us in front of the station and we were taken, in groups of eight, to a place nearby that turned out to be the base-camp of this day of expedition.

'At the base camp we had coffee and were welcomed by the managers of the project and introduced to the programme of the day (Figure 6). First we were introduced to the principles of human centred innovation and to a set of qualitative methods and techniques — various kinds of interviews and observations — all of which were inspired by the scientific field of ethnography and anthropology. After being divided into groups of eight we were introduced to our team-leader — who would lead us through the practical exercises of the day, an expedition to library user land (Figure 6).

Figure. 6

'My group was to meet users from one of the four categories, i.e., master's students, and we were going to do research at their workplace and their home. Master's students are ambitious people who are studying to promote their careers, they often have a family and thus their days are full of duties. How do libraries create services that are meaningful to this category of students?

'The first stop was at the workplace of a financial advisor, who gave us a guided tour around his office. In this case we were told to use the method of observation in work context, to observe and take notes about the physical items and atmosphere of the office, and to take photos — if allowed —, and we were. After the guided tour we conducted a basic interview based on open questions and genuine curiosity and we got the opportunity to ask clarifying questions.

'The second stop was intended to be at the home of a media consultant, another master's student. But since he worked in Copenhagen most days of the week, it had not been possible to arrange such a meeting. Instead he had been asked to provide a photo diary with pictures from his daily life and to write some text to each photo plus text that tied the photos together. After having examined the photo diary we did an interview with him in Copenhagen via mobile phone.

'Back at the base-camp we all had to categorize the data collected during observations and interviews (printed photos, video clips, observation notes, interview notes, and, in the case of taped interviews: transcripts, collected thoughts), by categorizing each segment under broader headlines. This phase was very crucial, we learned, as it is the careful preparation of data that lead to the final stage of the process: the analysis. Both phases should ideally be as broad and open as possible and conclusions not too easily and quickly drawn.

'When presenting our results, our group ended up by formulating three working questions, relevant for the creation of services for this category of users:

1) How do we relieve the master's students of their bad conscience?

2) How can libraries support talented employees to qualify for taking steps up the career ladder?

3) How can libraries work with information and services so that master's students' needs are met?

'These questions reflect the insights gained by actual meetings with the two master's students as ordinary people living ordinary lives — and not exclusively as library users.

'The Discovery Journey in Danish Library User Land thus gave me a range of concrete user-centred tools to choose from when I want to learn more about the needs of the users of my library. The methods and techniques are applicable in smaller as well as bigger library settings.' (Figure 7)

Figure. 7

The five other groups in the Aarhus expedition investigated users representing the three other categories and had the opportunity to try the other user-centred tools presented by the project.

Observations:

- Photo diary (described in text above)
- Observations in work context (described in text above)
- Think-aloud observations: library users are asked to solve certain problems while they think aloud during the execution of the assignment
-

Interviews:

- Basic interviews (described in text above)
- Value wheel: library users are interviewed in order to draw a picture of the extent to which their needs within a certain area are satisfied
- Interview at a location that means something to the user: library staff meet with library users on their home ground
- Small spontaneous interviews with users.
- Variations between interviews and observations:
- The meaning of physical surroundings: typical artifacts can be signs, boxes, furniture, doors, etc. — but may also include the way the light is set, the sound in a room and where it comes from, décor and much more
- Meaning of language — listening for metaphors: a useful tool to sharpen listening skills is to listen for the language. What are the terms and the metaphors used? Language tells a lot about the interviewed or observed person's approach to the world (Harbo and Biegel, 2010, p. 22 ff).

6. SEVEN PRINCIPLES OF HUMAN-CENTRED INNOVATION IN RELATION TO LIBRARIES

Many user investigations are carried out by using quantitative methods, collecting data through thoroughly designed questionnaires in order to get comparable data sets to analyze. However, quantitative investigations require big sets of data in order to be generally representative. Qualitative methods also require thorough preparation, but they can work with a smaller population of users, as qualitative methods do not claim generally representative results, but only results that are representative of the smaller categories of users actually investigated.

The methods used in this project belong to the latter category of user investigations, i.e., qualitative methods. Drawing on inspiration from the scientific field of ethnography and anthropology the project introduced the actual practicability of a set of abstract methods: a range of user-centred tools which could realistically be implemented even in small library environments — a criterion of success which had been defined by the project.

Teaming up with 'Changepilot' gave us a methodological basis for an open-minded, anthropological observing and listening approach. To make the approach practical in our everyday library work, 'Changepilot' formulated seven qualitative methodological principles about human-centred innovation which enable us to focus on user logic (Harbo and Biegel, 2010, p. 8 ff).

A short description:

1) Users run nothing

2) We prefer to use the term 'human-centred innovation' instead of 'user-driven innovation'. For most people it is difficult to put new words to what they need. Instead we try to uncover what drives people, what they dream of and how they live their lives.

3) Remove the blinkers

4) In areas where we basically are the experts (e.g., the library) our own knowledge may be the biggest obstacle that keeps us from seeing new angles. We must mentally remove our blinkers and put on new glasses to see things with new eyes.

5) Focus on the question

6) When working on human development one of the pitfalls is that we begin to search for solutions at the same moment we see a challenge. It is important to focus on the questions and to refine the question or to find more questions in order to be sure that we have formulated a question which is important for the users.

7) Meet ordinary people

8) We must meet and know ordinary people to find out what really matters to them to come up with exceptional solutions.

9) Focus on context

10) The decision to go to the library is taken before the visit takes place and the value of the visit is realized afterwards in the user's everyday life. Therefore we need to focus on the whole context surrounding the visit, including what happens before and after the visit.

11) Improve people's lives

12) We develop and innovate to improve people's lives. Nothing less!

13) Don't go to the zoo, go to the jungle

Human-centred innovation is forcing us out there where we meet the user in his or her own element, not in ours!

What does that mean for the librarian at work? Librarians who like to be in control of a situation and find proper answers are challenged by the human-centred approach to be open-minded and truly observe the user in his or her needs. The seven principles give answers to how and why to use this approach. How: principle 2, 3, 4, 5 and 7. Why: principle 1 and 6.

Perhaps the librarian experiences methodological pressure when using this approach. In that case the librarian must ask herself what is practically possible within her working context. Every new insight is valuable and is good training for the librarian using anthropological methods. So the use of the seven principles is scalable in relation to the librarian's working conditions. It should be possible to accomplish human-centred innovation if you are the only librarian in a small library or if you are a member of a team of librarians at a large university library.

At the same time living up to the seven principles keeps the 'anthropological librarian' on the right human-centred track.

'Users run nothing' means that the librarian should not only ask the user 'what do you need?', but observe and perhaps instead ask 'what is important for you?' in relation to the academic task, and in relation to the user's study life in general.

'Remove the blinkers' is very difficult for a librarian or any professional exploring his own area, but patience and not coming up with quick answers are ways to start seeing new things. A lot of our professional work is based on routine and common library knowledge, but trying to be a student or a researcher could provide us with new glasses.

In practice the librarian should dare to be an 'anthropologist' and go outside the library to follow a researcher or a group of students, e.g., by being introduced to the researcher's work process and teaching preparation, or by following study group meetings and courses, or lunch breaks by students.

'Focus on the question' is again a challenge to a librarian's need for control and the wish to give proper answers quickly. Normally a user will ask and the librarian will come up with an answer. Changing the roles will be difficult.

'Meet ordinary people' should not be a challenge to a librarian working with front end services. However, it is important to understand users not only as library users.

'Improve people's lives' could be specified as a university library's objective to improve not only education and research but also the lives of students and researchers.

'Don't go to the zoo, go to the jungle' means meeting the users on their home ground, where they work, study, have lunch and coffee breaks, live.

All in all human-centred innovation means a more proactive, process-oriented, observing and anthropological type of librarian who is ready to seek meaning and whose objective is to improve users' lives. The whole 'field trip to user land'-concept is a step towards giving librarians the competences to do that.

However, keep in mind that the concept can be implemented on a small scale as well as on a large scale, and that a pragmatic attitude is necessary.

7. CONCLUSIONS, SUSTAINABILITY AND FURTHER PERSPECTIVES

The DEFF project A Journey of Discovery in Danish Library User Land demonstrated concrete approaches and methods which in the future will make Danish librarians capable of perceiving library users in a new light. To perceive the library user as a person with a complex set of needs is very different from the way in which libraries have traditionally perceived their users. Complexity is here an important and meaningful element and insights in users' own understanding of their lives can contribute to radical changes in the way in which library services are presented.

However, the acknowledgement of the fact that the interpretation of users' behaviour is not necessarily the same as users' own interpretations is just as important. In this context the user-driven concept is substituted by the user-centred concept, which allows for inclusion of libraries' analysis of the field work. By implementing the range of user-centred tools presented by the project, libraries have a real possibility to start transforming the mindsets of librarians with regard to library users.

The project revealed completely new ways for us, library staff, to meet library users. We realized that in their lives (representing the user logic), libraries did not mean as much as we, library staff, think. Thus, the library staff participating in the expedition obtained new and valuable insights into the land of library users. The qualitative methods and techniques that they have been introduced to by the guides of the expedition have proven to be very useful and easily transferable when the staff returned to their own libraries.

8. EXAMPLES OF NEW WAYS OF LIBRARIANS MEETING LIBRARY USERS — INSPIRED BY THE PROJECT

- Observation of behaviour of patrons using a web application which tells them where a book should be placed in order to test if the application worked well in a certain context. The point was that saying is one thing, doing is another! This is why observation was preferred to interviewing.

- A small exploration of the understanding of the words 'community' and 'motivation' in relation to the question 'what does a good day of study look like'? University students at the Department of Social Sciences were asked this question by librarians using tools from 'A Field Trip to Library User Land'; they focused especially on the 'value wheel'.

- Another small investigation of the reason why some library users hesitate to ask the librarian/information counselor for help or, alternatively, started the conversation with 'I am sorry to disturb you'. Library staff were worried that they looked too busy, unfriendly or unapproachable. They observed library users and during one day they interviewed library users who started their approach with 'I am sorry to disturb'. Seven small interviews were done.

- A library team wanted to explore the concept of service by observing staff and customer interactions and by interviewing people using vox pop at 1) a combined swimming pool and library, 2) a shopping mall, 3) outside a tea and coffee shop, 4) a department store, 5) a MacDonald's, and 6) Ikea. What should their library service look like?

- Observations of library users combined with questionnaires sent to 180 students asking them whether the library should keep the collection of reference books or replace it with magazines about art, literature and music.

9. FURTHER PERSPECTIVES

Do we always want a library to be based on user logic (process oriented) innovation? We think it will make a lot of sense considering the fact that library services are targeted at users. Examples are the library catalogue, the classification system, the design of the library, the position of meetings point, information counseling services, etc.

We should also ask ourselves in which situations we prefer library logic (product oriented) innovation. In some cases, it is necessary to get a specific message through, e.g., how to navigate in a specific database.

According to the situation, either orientation can attribute to improve library users' lives.

Every librarian should be able to implement human-centred innovation in her library. The challenge for us is to further a human-centred innovation culture among librarians and to spread the word that it can be done in many different shapes and sizes. In our pragmatic understanding it is better to focus on the outcome than necessarily having one hundred per cent scientifically based evidence. If we succeed, we believe libraries and librarians will become flexible in adjusting to their users' needs.

With the easy access to information through, e.g., Google, Wikipedia and other websites the users' need for libraries providing access to information is perhaps decreasing. On the other hand the need of information counseling is increasing as access to enormous quantities of information resources requires information-literate users. Information counseling of users requires understanding of the context in which the users find themselves. The counseling librarian needs to know as much about the users' way of working, studying and living, as he or she knows about running a library in an efficient way.

One way of getting to know the users' needs is by applying the methods and techniques demonstrated in the project 'A Journey of Discovery in Danish Library User Land'. We recommend that a shift of focus from library logic to user logic be discussed in library environments as well as at academies of library and information science.

The project 'A Journey of Discovery in Danish Library User Land', its workshop concept, its how-to-do-manual containing concrete tools, and its website to share experiences and discussions, give us a practical fundament. Now it is up to us to do it!

REFERENCES

1. Kuhlthau, Carol C. (2004): Seeking meaning — a process approach to library and information services, Libraries Unlimited.

2. Harbo, Karen and Biegel, Rebekka (2010): A Field Trip to Library User Land. Deff — Denmark's Electronic Research Library and Changepilot

3. Deff — Denmark's Electronic Research Library: www.deff.dk

4. Changepilot: www.changepilot.dk

Chapter 7

Supporting Teaching and Research in an Online Environment: Developing the University of London Library Model

Sally Chambers

Supporting Teaching and Research in an Online Environment: Developing the University of London Library Model by SALLY CHAMBERS Founded in 1837, the University of London Library (ULL) is the central research library of the University of London with particular research strengths across the broader arts, humanities and social sciences. With the development of information and communications technology, it has become a hybrid library, offering access to electronic collections but also extending access to services and its physical collections through its website and online catalogue. Research libraries now face the challenge of supporting distance learning. The ULL has risen to this challenge by being a partner in the development of webenhanced distance learning by the University of London External Programme's Virtual Campus Project. To do this, the ULL has initiated the Virtual Library Service (VLS) Project. Within the research-focussed University of London, the ULL has anticipated the potential research-support needs of its students. It is also leading a separately funded Virtual Research Environment (VRE) Project on behalf of the University Libraries Committee to address these needs and inform the University of London's activities in the virtual sphere. The University of London has created the Electronic Library Projects Team (ELPT) to undertake the VLS and VRE projects and ensure that they develop along complementary lines. This paper outlines how the ULL is supporting teaching and research in an online environment through the development of its VLS and VRE models.

1. INFORMATION RESOURCES FOR DISTANCE LEARNERS

The initial project undertaken by the ELPT is to develop a dedicated VLS model to support the information needs of students registered with the University of London's External Programme, many of whom are studying at a considerable distance from London. Once fully developed, the VLS will offer access to a range of information and learning resources and related services as one of the principal elements in the External Programme's Virtual Campus Project.

2. THE UNIVERSITY OF LONDON EXTERNAL PROGRAMME

The University of London, established by Royal Charter in 1836, is one of the oldest universities in the United Kingdom. It is now a federation of seventeen self-governing colleges, all of which are members of the University by choice and are empowered to award the University of London degree. The University of London External Programme was established in 1858 to make University of London degrees accessible to students who were unable to come to the University to study. The degrees awarded to students on the External Programme are of the same academic standard as those awarded internally to students studying at one of the University's Colleges in London. Currently the Programme offers over 100 qualifications at undergraduate and postgraduate level to over 28,000 students in more than 180 countries. At undergraduate level over 65% of students are currently studying in the Asia Pacific region, whereas at postgraduate level, the students are more widely distributed, with 30% in the Asia Pacific, 30% in Western Europe and almost 30% from Africa and America combined. The University of London External Programme is a distinctly global programme with the majority of students studying at a considerable distance from London. Each qualification is led by a Lead College within the University, co-ordinated by the Lead Colleges Committee. Funded wholly from student fees, surpluses are reinvested in the Programme through new course developments and through major initiatives such as the Virtual Campus Project.

3. THE VIRTUAL CAMPUS PROJECT

In October 2000, the External Programme embarked on a period of significant investment to develop new and redevelop existing qualifications, the majority of which will be delivered wholly or partially online. In parallel with these academic developments, the External Programme is currently developing an online environment, the Virtual Campus. Anyone visiting the Virtual Campus from a computer anywhere in the world will be able to access a range of information, advisory, resource and support services as would be expected on a physical campus mirrored into a virtual context. The Virtual Campus Project is a collaborative initiative bringing together three central academic services of the University of London: The External Programme, The ULL and the University of London Computer Centre. Running from October 2000 to July 2003, the aim of the project is to provide a common infrastructure to support the development, delivery and management of internet-based courses and services for the External Programme.

4. THE VIRTUAL LIBRARY SERVICE

One of the central services within the Virtual Campus Project is the VLS, which is being co-ordinated by the ELPT. Running from October 2000 until July 2003, the aim of the VLS project is to: „provide access to a range of learning and information resources and services for students registered with the University of London's External Programme from wherever they have chosen to study." Primarily, the VLS will mirror the core resources and services found in a university library, however, as more courses are delivered online, the integration of library resources within Virtual Learning Environments (VLEs) will be explored. As well as providing learning and information services for the lifetime of the Virtual Campus Project, the VLS Project will develop a model for the delivery of a sustainable service. Appropriate definitions of service will be created, including recommendations of how such sustainability may be achieved.

The VLS provides a variety of online resources and services put together in liaison with each course development team participating in the Virtual Campus Project. The principal components in the VLS are:

- Online resources: electronic datasets and internet resources targeted at each course within the Virtual Campus Project. There will also be a reference collection enabling students to find rapid reference facts to support their studies.

- Enquiry service. Available by telephone, email or web form.

- Information literacy skills. Externally developed guides and tutorials are a part of the initial VLS. The VLS project will also provide quick start guides to the resources it offers and aims to develop its own tutorials tailored to suit the specific needs of the External Programme.

- Authentication. Exploring various methods of authenticating users will be of high priority to ensure a seamless experience within the Virtual Campus.

- Web-based portal with tailored entry points for each External Programme course.

5. IMPLEMENTATION

Online Resources: Acquisition and Integration One of the most significant tasks in the development of the VLS has been working closely with the academic programme development teams to determine which resources will best support the delivery of their courses. Initially, during the lifetime of the project, the VLS will provide library support for approximately fifteen courses across a wide variety of subjects including business, dentistry, economics, education, finance, healthcare, history and law. Some of the resources will be specialised, such as those for dentistry, and will only be applicable for one programme. However, other resources may be applicable to several programmes, due to the overlap between the disciplines, such as in the areas of business, finance and economics. It has been necessary to maintain a map of the resource requirements across the entire project to ensure that they are most cost-effectively acquired. In addition to paid-for resources, which so far mainly focus on subject specific databases and e-journals, the academic programme development teams have been keen to provide relevant websites and subject gateways, which are being collected and organised as part of the VLS.

The increase in both electronic resources and the online course delivery is providing the opportunity for a closer relationship between library resources and academic content. Such integration increases the necessity of the library and academic staff working closely together throughout course development. In the UK, nationally funded projects such as INSPIRAL (INveStigating Portals for Information Resources And

Learning)1 and ANGEL (Authenticated Networked Guided Environment for Learning)2 , are already working closely on how such integration can be best achieved.

6. SUPPORT SERVICES: QUICK START GUIDES AND THE ENQUIRY SERVICE

Support for the use of the resources in the VLS is essential. Many of the External Programme students are studying both on their own and at a distance and may not be familiar with online study and electronic resources. Information about the resources and services offered is sent to each student who has access to the VLS, including Quick Start Guides explaining the basic functions of each the resources. However, it is important that an opportunity to ask questions is made available in addition to the information they receive.

The VLS enquiry service was launched in September 2001 to support the resources for the Laws Programme. Enquiries are answered, within two working days of receipt, by the Electronic Library Projects Assistant with the support of an experienced team of information and subject staff. Having a main contact person for the VLS will help to reduce the isolation experienced by students who are studying at a distance, as they will regularly contact the same person. In addition, contact details and areas of responsibility for the whole ELPT, including photographs, will be added to the VLS website to further increase the personal tone of the service. Although this is a virtual library service, it is important that students can choose the way they prefer to communicate with the library; either by email, telephone, fax, letter or perhaps even face to face if possible. Furthermore, in order to provide a dedicated service, a separate telephone number has been set up for enquiries from External Programme students. In the first instance queries are only answered between 09.00 and 17.00 (Monday to Friday) UK Time. However, due to the range of time zones in which External Programme students are studying a voice mailbox has been set up so that the library can receive telephone enquiries, as with email, 24 hours a day 7 days a week. Questions received via the VLS are recorded and analysed in order to influence the future development of the enquiry service. Already, an electronic help desk has been developed which provides answers to Frequently Asked Questions (FAQs) that have been received through the VLS enquiry service. Students are advised to consult this helpdesk before they ask a question. Since its launch, usage of the enquiry service is increasing clearly demonstrating that the students need it.

7. AUTHENTICATION

Once a student has logged into the Virtual Campus she should not be aware that any other authentication is taking place; she should not have to enter her username and password again. Any further authentication required, such as for electronic library resources, will be done behind the scenes using data exported from the External Programme's student registry. In the case of the VLS, it is essential that only authorised students have access to the resources to ensure compliance with licensing agreements. Ultimately, the VLS will provide access to a comprehensive range of resources, from a wide variety of suppliers, many of which will have different methods of provided authenticated access to their resources. A service-wide authentication is therefore a necessary component in the delivery of the service.

In the UK, the problem of multiple passwords for electronic library resources has been to some extent bridged by the development of the Athens3 , an access management system for UK Higher and Further Education. Athens provides off-campus access to a range of electronic library resources with the input of a single username and password, however, at this stage, not all resource providers will allow their resources to be authenticated by Athens. In addition, it is also not possible to integrate Athens with the authentication system that is being used to provide access to the VLEs as part of the main Virtual

Campus Project. In this way, currently students are not able to login to the Virtual Campus and have access to all their resources and service without having to type in additional usernames and passwords.

For on-campus behind the scenes or quiet authentication4 access to resources is often provided by Internet Protocol (IP) recognition. For this method, the library registers the IP addresses of all the library computer terminals with the resource provider. When a student tries to access the resource from a computer with a registered IP address then they get immediate access to the resources without the necessity for any further authentication. However, for a VLS, with potential 28,000 students, it would be impractical, and not secure, to register the IP addresses of all the students. Resource providers initially have provided remote access to such resource by authenticating students by username and password, however this will not create the seamless experience we are aiming for.

The ELPT is currently investigating a variety of authentication solutions including the use of proxy servers, which work in a similar way to IP address recognition, but without the need for the registration of student IP addresses. Attention is also being paid to the development of a second generation of Athens access management system5 . As Athens is a national access management system and is often seen as a de facto authentication standard, its use as the ultimate authentication solution for the VLS, if possible, would be preferable. However, while this investigation is in progress, most of the VLS' resources will be authenticated via the current Athens with the input of a single username and password wherever possible. In this way, fairly quiet authentication can be provided.

8. INTEGRATING ACCESS TO THE VIRTUAL LIBRARY SERVICE

The aim of the VLS is to provide seamless access to electronic library resources and services as part of the online courses. The VLS is also a coherent website providing access to electronic library resources and services independently from any online courses. When designing the VLS, a flexible infrastructure was therefore needed to satisfy two seemingly incompatible criteria. The prototype VLS was launched in September 2001. At this stage, the online course developments were still in their infancy and therefore the VLEs were not available for testing. It was therefore important to keep the design of the VLS website as open as possible to allow the pages to be integrated within the various VLEs once they had been developed. In order to increase the flexibility of the website it was decided that it should be partially database driven. Constructed in Access, the database contains the URLs (Uniform Resource Locators) for all the resources provided in the VLS. The webpages are published dynamically, using a combination of SQL (Structured Query Language) queries and ASP (Active Server Pages). This solution allows the same resource to appear on different webpages from a single URL within the Access database. The resources can therefore appear on a page within the coherent VLS website, but at the same time appear on a separate resources page within an online course. By Summer 2002, several VLE prototypes will have been developed and will therefore be available to prototype the integration of the library resources. In addition, the Electronic Library Projects Assistant will carry out an evaluation of the current site in terms of standards and good practice in web design and feedback from site users. This evaluation will take into account issues such as usability, accessibility, scalability, usage patterns and site maintenance and be used to further improve the current site. Wherever possible the feedback of the users of the VLS will input into the redesign of the website, which will be launched in time for the beginning of the new academic year in September 2002.

9. ISSUES FOR THE FUTURE

During the initial stages of the VLS project, considerable work has been achieved, particularly on the development and realisation of the VLS model. As 2003 approaches, the majority of the courses will be registering their first cohorts and so with the resources in place the real test of VLS will begin. With the VLS in place, appropriate definitions of service will be created, including recommendations of how a sustainable service may be delivered beyond the lifetime of the project.

10. SUPPORTING RESEARCH ONLINE

The Virtual Research Environment Project A complementary project undertaken by the ELPT is the Virtual Research Environment (VRE) project, supported by a grant from the Vice-Chancellor's Development Fund to the University Libraries Committee. Running from December 2001 to November 2002, the aim of the VRE project is to: „ develop and test an integrated web-based research environment to support the needs of the e-researcher." This project builds on the work that is currently done to provide online library services for the University of London External Programme's Virtual Campus Project, as outlined above.

Although considerable work is being undertaken in the field of online teaching and learning, online research support is currently relatively unexplored. As a research-led university, the University of London is well aware of this disparity and has begun to develop online support for both research elements of taught-course dissertations, in addition to support for research degrees online.

Technology already exists to support the taught elements of online degree courses, both at undergraduate and postgraduate level. VLEs may be used to support a range of learning contexts from conventional classroom delivery to distance and online learning. Such software has been evaluated through the Virtual Campus Project and elsewhere. However, in order to support academic research, online software (such as groupware) must be examined for its suitability in satisfying the needs of the e-researcher.

The VRE Project is grounded in a review of current research support provision within the academic community and an analysis of the needs of the research community within the University of London. Currently a user survey is being prepared with the aim of establishing which functionalities are needed in a VRE. As the strengths of the ULL are in the broader arts, humanities and social sciences, the VRE will focus on the needs of the research communities in these areas. However, the differing needs of researchers in the physical sciences will be explored in the Research Support Review.

11. PRINCIPLE COMPONENTS

The VRE will identify and integrate principle components required for effective support of online research. This will be informed by the research support review and needs analysis. However, it is anticipated that principle components of the VRE may include.

12. VIRTUAL RESEARCH LIBRARY SUPPORT

An integrated resources page within the VRE could provide software to store and compile the researcher's electronic bibliography. Personalised harvesting tools which can search both the internet and across various relevant library resources (e.g. e-journals, subscription services, even the library catalogue) on a regular basis using predetermined searches could also be included. The results of which could be added to the electronic bibliography. Furthermore, an online research skills tutorial which can be applied across a number of research scenarios could also be developed and included, for which a separate bid is being prepared.

Using the research proposal as a starting point, a subject specialist librarian from within the researchers subject area could provide a specialised gateway to relevant resources available through the library.

In addition research-related questions could be answered using an electronic enquiry service and online information skills training also be developed. These services demonstrate possible synergies between the VRE project and the services delivered as part of the VLS project described above.

Research-related Information

Provision of general web-based research-related reference information, including University Research Guidelines; Citation Style Manual; Research Methods; Research Funding Sources within and outside the University of London.

Online Secure Research Repository (OSRR)

A secure environment where elements of the research, (e.g. research proposal; literature review; methodology) can be shared between researcher and supervisor could be included. In this context the VRE could contribute to the electronic infrastructure of distance research degrees that are currently being discussed within the University of London.

Online Research Support Mechanisms

The system could provide mechanisms to allow tutors and students to interact in synchronous and asynchronous modes. These mechanisms could include: discussion boards supporting subject-based discussions; integrated electronic mail; video conferencing and real-time discussion. This could encourage the development of subject-specific research communities from across the University and the provision of online seminars and presentations.

Tracking of Research Activity and Achievement

Online research planning tools to track research activity and achievement could be included perhaps incorporating project management software.

Research Output Repository Research

Output Repository containing digitised or born-digital research output such as research articles, book reviews, electronic theses and dissertations could form a valuable element of the VRE.

Software Evaluation

In parallel with the identification of the principal components of the VRE, the evaluation of available software to form the VRE infrastructure is taking place. Currently a development environment to assess the suitability of Open Source software packages has been set up, although this does not necessarily anticipate the use of Open Source Software in the final implementation. It is anticipated that the principal components will be developed as modules that can easily be put together and customized by researchers. In this way it will be possible to set up different instances of the VRE for different subject areas. At present the first customized VRE-application is being set up in collaboration with the International Portal for Manuscript and Book Studies within ULL, a development of the RSLP funded project; Palaeography: Developing the National Resource. [6]

Researcher Involvement

The generic VRE model will be tested and evaluated with a VRE user group that is currently being formed from an appropriate sample of the University of London research community. This user group will have a role throughout the development of the VRE, such as in the identification of the principal com ponents, as well as in the testing and evaluation of the final model. A Strategic Advisory Group for the VRE project has also been set up to monitor and guide the progress of the VRE Project. The membership of the group is designed to ensure close harmonisation and exchange between the Virtual Campus Project, the Society of College, National and University Libraries' (SCONUL) E-University Task Force, the Consortium of University Research Libraries'

(CURL) Task Forces on Teaching and Learning and Scholarly Communications and other relevant initiatives.

13. CONCLUSION

The work of the ELPT will potentially influence the ULL's future support of teaching and research in an online environment. The projects outlined above provide a test-bed for the development of new services that may be integrated into ULL's central service delivery.

Chapter 8

Improving Library Management by Using Cost Analysis Tools: A Case Study for Cataloguing Processes

Lorena Siguenza-Guzman[1], Alexandra Van den Abbeele[2], Dirk Cattrysse[3]

[1]Department of Computer Science, Faculty of Engineering, University of Cuenca, Ecuador and Centre for Industrial Management Traffic and Infrastructure, Faculty of Engineering, KU Leuven, Belgium

[2]Faculty of Business and Economics, KU Leuven, Belgium

[3]Centre for Industrial Management Traffic & Infrastructure, Faculty of Engineering, KU Leuven, Belgium

ABSTRACT

TTDABC is a relatively new costing management technique, initially developed for manufacturing processes, which is gaining attention in libraries. This is because TDABC is a fast and simple method that only requires two parameters, an estimation of time required to perform an activity and the unit cost per time of supplying capacity. A few case studies have been documented with regard to TDABC in libraries; all of them being oriented to analyse specific library activities such as inter-library loan, acquisition and circulation processes. The primary focus of this paper is to describe TDABC implementation in one of the most important library processes, namely cataloguing. In particular, original and copy cataloguing are analysed through a case study to demonstrate the applicability and usefulness of TDABC to perform cost analysis of cataloguing processes.

KEYWORDS

Costing analysis; Time-Driven Activity-Based Costing; Cataloguing Processes; Academic Library

1. INTRODUCTION

In the current economic situation, characterized by periodic shortages and limited budget resources, libraries are in search of methods to improve their process efficiency and provide high-quality services at lower costs (**ACRL Research Planning and Review Committee, 2010**; **Cottrell, 2012**). In order to improve their performance, library managers need to consider cost reduction and the inclusion of service costs in their decisions (**Hoozée, Vermeire, & Bruggeman, 2012**). Furthermore, they should strive to identify improvement opportunities and to eliminate costs related to non-value adding activities (**Ellis-Newman, Izan, & Robinson, 1996**). In order to do this, library managers must keep their activities, resources and costs under control, relying on valid information about activity costs, resource capacity and their performance (**Stouthuysen, Swiggers, Reheul, & Roodhooft, 2010**).

In service institutions such as libraries, several methodologies for costing analysis have been used for decades, whereas the Time-Driven Activity-Based Costing (TDABC) is one of the most recent approaches.

TDABC is a cost management technique developed by Kaplan and Anderson to overcome the difficulties presented by previous costing systems (**Kaplan & Anderson, 2007**). By implementing TDABC in libraries, key benefits are expected, including the possibility of benchmarking different scenarios; identifying non-value added activities; and justifying decisions and choices for staff recruitment, training and new service development (**Siguenza-Guzman, Van den Abbeele, Vandewalle, Verhaaren, & Cattrysse, 2013a**).

Although some research has been carried out with respect to TDABC in libraries, these studies have been focused on specific library activities such as the inter-library loan (ILL), acquisition, and circulation (**Siguenza-Guzman, Van den Abbeele, Vandewalle, Verhaaren, & Cattrysse, 2013b**). More research is still required to determine whether TDABC can be useful in other library services. Therefore, the aim of the paper is to provide more detailed insight on implementing TDABC for library cataloguing processes. We focus on this unit because cataloguing is considered for many libraries, to be one of their most expensive processes, especially original cataloguing (**Manaf & Rahman, 2006**). The remainder of this paper is organized as follows. Firstly, a brief outline of the theoretical background of costing systems is provided (§ 2). Secondly, the different steps involved in implementing TDABC in original and copy cataloguing are explained (§ 3). Thirdly, we discuss the similarities and differences found in the original and copy cataloguing regarding time and cost per activity (§ 4). In addition, a number of recommendations where process improvements were unveiled are included in this section. Fourthly, the benefits encountered when implementing the TDABC model in an academic library in Belgium are described (§ 5). We end this paper with a brief conclusion (§ 6).

2. THEORETICAL BACKGROUND: COSTING SYSTEMS

2.1. Traditional Costing Systems

Several library cost analysis studies have been performed since the 1970s (**Roberts, 2003**). However, these studies were mainly treated as technical rather than organizational or managerial innovations (**Kont, 2011**). **Ellis-Newman et al. (1996)** report that the majority of prior studies on library costs were undertaken in the United States, utilizing cost allocation models more compatible with traditional costing methods. Traditionally, the total product cost consists of direct costs, such as the cost of materials and direct labour, and a percentage of overheads as indirect costs (**Siguenza-Guzman et al., 2013a**). The latter includes training, marketing, and infrastructure, among others. Traditional costing systems are adequate when indirect expenses are low and product variety is limited. However, in an environment with a broad range of products and enhanced services, such as a library, indirect costs have become substantially more complex than direct costs (**Siguenza-Guzman et al., 2013a**). This situation renders traditional methods inadequate; not only for estimating the effect of strategic decisions, but also for providing crucial information to library managers (**Ellis-Newman & Robinson, 1998**; **Kaplan & Cooper, 1998**).

2.2. Activity-Based Costing Systems

Activity-Based Costing (ABC) is an advanced costing calculation that seeks to remedy the limitations of traditional methods (**Kaplan & Cooper, 1998**). ABC, promoted by Cooper and Kaplan in the mid-1980s, first accumulates indirect costs for each activity and then assigns the activity costs to the services causing that activity (**Cooper & Kaplan, 1988**; **Ellis-Newman & Robinson, 1998**). ABC has proven to be a valuable tool for libraries through its implementation in several case studies. For instance, **Ellis-Newman et al. (1996)** examined the application of ABC in academic libraries of two Western Australian universities. This study illustrates how activity costing helps managers to differentiate key activities from others that do not add value. In an additional study, **Ellis-Newman and Robinson (1998)** discuss the benefits of ABC for library managers and the steps involved in implementing ABC in an academic library. The authors show how traditional costing systems are unable to explain the relationship between costs and the underlying activities. Furthermore, **Ellis-Newman (2003)** demonstrates the type of information that an ABC system provides to assist decision-making with a case study in the user services area of an Australian academic library. In turn, **Ching, Leung, Fidow and Huang (2008)** employ ABC to examine the Super e-Book Consortium in Taiwan and Hong Kong. The study finds cost drivers of consortia business operations, and identifies the key consortium activities and their relevant costs. Moreover, **Goddard and Ooi (1998)** examined the development of ABC through a case study applied to library services at the University of Southampton. The authors present ABC as an option to overcome some of the problems of overhead allocations. Despite the benefits of ABC, they also describe significant problems with its practical application such as the amount of resources and time required for its development and maintenance. Eventually, **Novak, Paulus and Clair (2011)** describe how a medium-sized university library implemented ABC and other decision-making strategies to make budgetary cuts and thereby, redirecting library services.

Although a relatively extensive stream of literature finds that ABC systems provide interesting advantages for decision making in libraries, ABC also has its limitations. **Kaplan and Anderson (2004, 2007)** note that ABC is difficult and costly to implement and maintain, especially when the current accounting system does not support the collection of ABC information. Data collection is time consuming and costly because of the need to interview and survey the library staff to estimate the percentage of time spent on each activity. While it works well in small organizations and limited activities, it becomes problematic to scale up to larger organizations (**Hoozée et al., 2012**). Managers also question the accuracy of the system since cost assignments are based on individuals' subjective information on how they spend their time. Furthermore, staff resistance could arise, as employees might feel threatened by the suggestion that their work should be improved. As a consequence, ABC systems tend to be outdated, and in some cases are abandoned and substituted by less demanding approaches such as Time-Driven Activity-Based Costing (**Wegmann & Nozile, 2009**; **Yilmaz, 2008**).

2.3. Time-Driven Activity-Based Costing Systems

TDABC is a useful cost management technique developed by **Kaplan and Anderson in 2004** to overcome the difficulties presented by previous costing systems (**Kaplan & Anderson, 2004**). TDABC assigns resource costs directly to cost objects using a fast and simple framework that only requires the unit cost of supplying resource capacity, and an estimation of the time duration of an activity (**Kaplan & Anderson, 2007**). Unlike the percentages that employees subjectively estimate for an ABC model, the time duration in a TDABC model can be readily observed and validated (**Kaplan & Anderson, 2007**). For each activity, «costing equations» are calculated and computed by time equations, which are the sum of individual activity times (**Yilmaz, 2008**). Through the use of time equations, TDABC allows incorporation of variation in the time demands made by different types of transactions, and consequently the representation of all possible combinations of activities that a process performs (**Kaplan & Anderson, 2007**). Five main TDABC advantages, highlighted by **Siguenza-Guzman et al. (2013a)**, are its simplicity to build accurate models and improve the understanding of the different processes; the opportunity of modelling complex operations thanks to the use of multiple drivers; a good estimation of resource consumption and capacity utilization; its fast maintenance compared to ABC models; and the possibility of using TDABC in a predictive manner.

TDABC has been carried out in specific library activities such as inter-library loan (ILL), acquisition and circulation. For instance, the first case study by**Pernot, Roothooft and Van den Abbeele (2007)** uses TDABC to calculate ILL costs and describes TDABC as a useful technique to reduce ILL resource costs and to renegotiate ILL service prices based on more accurate costs. The authors conclude that TDABC is very suited to cope with increasing cost pressures, and its findings can contribute to improve library services at lower costs. A second case study presented by **Stouthuysen et al. (2010)** describes the use of TDABC for a library acquisition process. The authors state that TDABC provides library managers with a better insight into cost drivers, visualizes the acquisition process efficiencies and capacity utilization, and leads to potential cost efficiencies. As an illustration, they consider that 50% of some costs could be saved if administrative assistants get involved in the acquisition process instead of being performed by the head of department, provided that they are capable of doing these tasks. In addition, the authors demonstrate that TDABC can be updated rapidly and inexpensively to changes. Due to this flexibility, they consider that TDABC can be applied to other processes such as cataloguing or digitalized activities, with significant benefits. The latest case study by **Siguenza-Guzman, Van den Abbeele, Vandewalle, Verhaaren and Cattrysse (2014)** uses TDABC to analyse lending and returning processes. The authors provide several important insights for a successful implementation of TDABC in libraries such as: 1) collecting the time duration of the activities through direct observation to improve the level of accuracy; 2) using graphical representations of activity flows to validate the collected information straightforwardly; and 3) showing that clarifying the measurement purpose is crucial in improving the level of acceptance, and achieving the desired commitment with staff. They conclude that the TDABC implementation is worthwhile since it leads to a more accurate cost and process analysis for supporting decision-making.

3. TDABC IN CATALOGUING PROCESSES

The data used for this research was collected at the Arenberg Campus Library1(Campusbibliotheek Arenberg, hence the abbreviation CBA) of KU Leuven in Belgium. CBA offers information sources on subjects of the exact sciences, engineering, architecture, kinesiology, and rehabilitation sciences (**Campus Bibliotheek Arenberg, 2013**). Its services are handled by approximately 20.5 full-time equivalent employees (FTE). In this case study, we focus on describing the application of TDABC in two types of cataloguing activities: original and copy cataloguing. The former refers to creating a new bibliographic record from scratch, while the latter to adapting a pre-existing record to the characteristics of the item in hand (**Reitz, 2004**).

For this case study, qualitative interview data with quantitative data analysis were combined following the six steps presented by **Everaert, Bruggeman, Sarens, Anderson and Levant (2008)** to calculate the cost of activities through the TDABC model. These steps illustrated in Table 1 are described in detail by**Siguenza-Guzman et al. (2014)**.

Table 1: Time-Driven Activity-Based Costing steps (**Everaert et al., 2008**).

Time-Driven Activity-Based Costing	
Step 1	Identification of resource groups
Step 2	Estimation of the total cost of each resource group
Step 3	Estimation of the practical capacity of each resource group
Step 4	Calculation of the unit cost of each resource group
Step 5	Estimation of the standard time duration of each activity
Step 6	Multiplying the unit cost of each resource group by the time duration per activity

To identify resource groups involved in cataloguing activities as required in Step 1, multiple interviews were conducted. Initial interviews started through brief discussions with the library manager, and then moved to a more detailed level with the library staff. For each activity, a final interview was performed in order to validate specific details about the different sub-activities. During the interviews, key activities involved in each process were described in detail by employees in charge. The information was used to build flow charts of activity sequences. As **Siguenza-Guzman et al. (2014)** indicate, flowcharts allow a good overview of the different activities performed in a process, to identify additional expenditures such as computer maintenance and software licences, and afterwards to validate the activities in an optimal and simple manner. Figure 1shows the activity flow of original and copy cataloguing respectively.

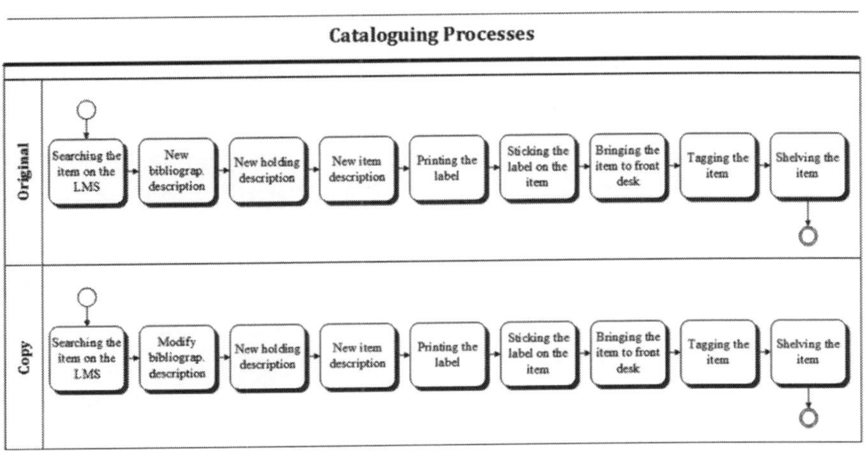

Note. – LMS = Library Management System.

Figure 1. Original cataloguing process.

The total cost of each resource group required in Step 2 was provided by the accountant and library manager via the Library Management System (LMS). The costs were classified into direct and indirect costs. Direct costs included salaries of staff and student library employees —

SLE (i.e. students hired to perform secondary activities), equipment, and technology. Conversely, examples of indirect costs included stationery, electricity, support, telephone, training, and other items used to perform an activity (**Vazakidis & Karagiannis, 2009**).

The salary of cataloguers were calculated based on the average salary earned by employees responsible for cataloguing. According to the Chief Librarian, the total number of personnel assigned to cataloguing represents 1 full-time equivalent (FTE). The 1 FTE consisted of six people, each dedicating different amounts of their time to cataloguing processes. This corresponded to about €59,000 on a yearly basis. LMS costs covered the annual integrated library system and supporting software license fees attributed for the CBA. The Library Management System includes functionalities for acquisition, cataloguing, circulation and reporting. This integrated library solution was acquired for the entire library which consisted of the Central Library, three Campus Libraries and several Faculty Libraries for the Humanities and Social Sciences group. The LMS costs amounted to €17,000 on a yearly base. The annual computer maintenance costs for specific tasks such as reparation, maintenance, cleaning and depreciation of a PC in the cataloguing processes equaled €5,000. RFID maintenance costs refer to the costs associated with the maintenance, repair and inspection of the RFID system and yearly costs corresponded to about €17,000.

In the case of indirect costs, the library accountant estimated that about €195,000 was annually spent on general overhead (GO) costs. GO costs included management, secretary, accounting, training, staff meetings and stationery material. Other indirect values such as electricity, telephone, heating and transportation were not accounted as part of general overhead costs since they were paid by the University and not charged to the library (**Ellis-Newman & Robinson, 1998**). In order to calculate the overhead costs attributed to cataloguing activities, general overhead cost was divided by the total number of FTE working at the entire library. This resulted in a yearly overhead of approximately

€9,500 per FTE. An overview of the total cost of each resource group can be seen in Table 2.

Table 2. Total cost of each resource group.

Total cost per resource group	
Resource Group	**Cost (€) per year**
Cataloguers	€59,000
Library Management System	€17,000
Computer Maintenance	€5,000
RFID Maintenance	€17,000
General Overhead	€9,500

Then, the practical capacity estimation of each resource group was calculated in Step 3. According to **Kaplan and Anderson (2007**, pp. 52–53), practical time capacity can be estimated in two different ways: 1) assuming an 80% of theoretical time capacity for people due to breaks, arrival and departure, training, meetings, and chitchats; and 85% for machines due to maintenance, repair, and scheduling fluctuations. 2) Calculating the real values according to the library situation, for example, available working hours, excluding holidays, meetings and training hours. In order to simplify the study, the first option was selected. For staff capacity, 38 hours per week were accounted as theoretical time capacity. It means 30.4 hours per week for practical capacity (staff practical capacity = 38 hours * 80%). Assuming fifty-two weeks per year, the practical capacity of a cataloguer is equal to 94,848 minutes per year $\left(30.4\frac{\text{hours}}{\text{week}}\times 52\frac{\text{weeks}}{\text{year}}\times 60\frac{\text{minutes}}{\text{hour}}\right)$. In the case of machines, the theoretical time capacity was set equal to the time in which cataloguers were available that is again 38 hours per week. Thus, the practical capacity for machines is 32.3 hours per week (machines practical capacity = 38 hours * 85%), and 100,776 minutes per year $\left(32.3\frac{\text{hours}}{\text{week}}\times 52\frac{\text{weeks}}{\text{year}}\times 60\frac{\text{minutes}}{\text{hour}}\right)$.

Once the practical capacity was obtained, the cost per unit time was calculated in Step 4 by dividing the total cost of a resource (Step 2) by the practical time capacity (Step 3). An overview of the resulting costs involved in this analysis is shown in Table 3.

$$\text{Cost per unit time} = \frac{\text{time cost of the resource}}{\text{practical capacity}}$$

Table 3. Costs involved in the analysis.

Total cost per minute for the different resource groups	
Resource Group	Cost per minute (€/min)
Cataloguers	0.62
Library Management System	0.17
Computer Maintenance	0.05
RFID Maintenance	0.17
General Overhead	0.10

For the fifth step, **Kaplan and Anderson (2007**, p. 30) recommend that the time required to perform an activity should be estimated based on standard times rather than actual times, since these times might reflect random variation, individual employee variation and nonrecurring factors. Moreover, the authors argue that precision is not critical and that a rough accuracy is sufficient because gross inaccuracies will be revealed either in unexpected surpluses or shortages of committed resources. This level of accuracy can be obtained by multiple methods such as direct observation, interviews, process maps or leveraging time estimates from elsewhere in the institution (**Kaplan & Anderson, 2007**, p. 26). In our case study, the standard time to perform an activity was gathered in the academic period 2010–2011 through direct observation as recommended by**Siguenza-Guzman et al. (2014)**. Observations were made multiple times using a stopwatch during several days at different hours in order to avoid possible biases (**Siguenza-Guzman et al., 2014**). Finally, based on their average values, an additional interview was performed in order to validate the data collection. Next, time equations were constructed for each activity. Time equations are the sum of individual activity times, which are represented with the following expression (**Kaplan & Anderson, 2007**):

$$\text{Time required to perform an activity} = (\beta_0 + \beta_1 X_1 + \beta_2 X_2 + \beta_3 X_3 + \beta_4 X_4 + \beta_5 X_5 + \ldots + \beta_i X_i)$$

With:

β_0= The standard time to perform the basic activity (e.g. 2 minutes)

β_i= The estimated time for the incremental activity *i* (e.g. time required for a librarian to enter an item in the cataloguing system = 0.5 minutes)

$\mathbf{X_i}$= The quantity of incremental activity *i* (e.g. items per batch = 1, 2...)

3.1. Original Cataloguing

The process, as shown in Figure 1, starts by searching the item in hand on the LMS in order to verify whether a similar record is already present in the database. This searching takes on average of 57s. If the item and record do not appear to match, the cataloguer creates a new record which includes the bibliographic description, requiring 306s (or 5min and 6s). A bibliographic description is the standardized description of an item including: title, edition, material specific details, details of publication, standard number, etc. (**Reitz, 2004**). The cataloguer then creates a new holding description, which is usually the information concerning the location of an item, taking 32s; and finally a new item description in 15s. Item description indicates item type, volume number, barcode, and loan rules. Once the new record is processed and stored in the database, the cataloguer prints the corresponding label (30s), and sticks on the item in 28s. Afterwards, the cataloguer brings the item to the front desk (30s) in order to tag the item (40s). An individual tag costs €0.30 including VAT. Eventually, the item is placed on the corresponding shelf or stack, requiring 168s (or 2min and 48s).

3.2. Copy Cataloguing

In contrast to the above process as shown in Figure 2, the cataloguer does find a record that appears to match with the item in hand. The cataloguer requires 74s (or 1min and 14s) to validate and modify the bibliographic description. Next, the cataloguer creates a new holding and item description, taking 32s and 15s respectively. Labelling and shelving are the same as the original cataloguing process.

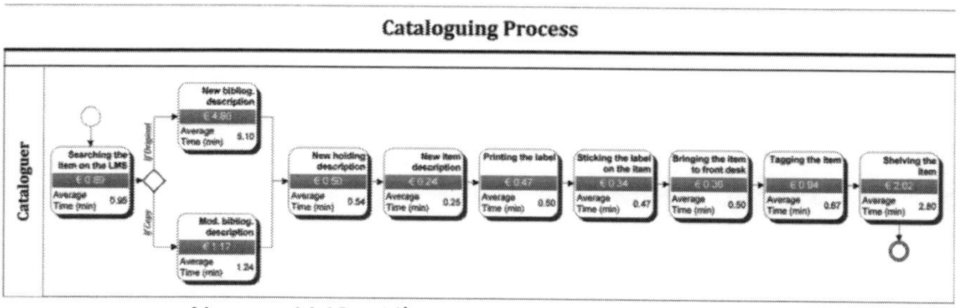

Note. – LMS = Library Management System.

Figure 2. Cataloguing process.

As the only difference between original and copy cataloguing is new versus modifying bibliographic description, we can create only one time equation by adding dummy variables in the equation. A dummy is a variable that takes the value 1 or 0 if a certain condition is true or false respectively. The resulting equation is as follows:

$$\begin{aligned} \textit{Cataloguing} = {} & \text{Searching} + \text{New}_{\text{Bibdescription}}\{\textit{if original\_cataloguing}\} \\ & + \text{Mod}_{\text{Bibdescription}}\{\textit{if copy\_cataloguing}\} + \text{New}_{\text{holding}} + \text{New}_{\text{item}} + \text{Print}_{\text{label}} \\ & + \text{Stick}_{\text{label}} + \text{Bring}_{\text{item}} + \text{Tag}_{\text{item}} + \text{Shelve}_{\text{item}} \\ \textit{Cataloguing} = {} & 0.95 + 5.10\,\{\textit{if original\_cataloguing}\} + 1.24\{\textit{if copy\_cataloguing}\} \\ & + 0.54 + 0.25 + 0.50 + 0.47 + 0.50 + 0.67 + 2.80 \end{aligned}$$

Once the estimated time per activity and the unit cost of each resource group is calculated, costs are assigned to cost objects by multiplying the unit cost per time of resources by the estimated time required to perform the activity. It is represented by the following mathematical expression:

*Cost of an individual activity = time required to perform an activity * resources cost*

Figure 2 shows the resulting activity flow of the cataloguing process including average times and costs per sub-activity. Original and copy cataloguing are integrated in the same graphical representation by using decision diamonds. Finally, the total cost of the process is calculated by summing up all activity costs and can be represented by:

$$\textit{Total Cost of a Process} = \sum \textit{costs of individual activities}$$

The total costs and time incurred in the original and copy cataloguing processes are presented in Table 4. The resulting table is divided vertically into six columns and horizontally by standard and optional activities to separate the activities influenced by dummy variables. The first column lists the activities identified in the cataloguing process; the second column shows the average time per activity. The third column specifies the accumulated costs per minute of resources involved in each activity; the fourth column calculates the resulting cost incurred in the activity. The fifth column describes the condition under which two options are available to be selected: original and copy cataloguing.

Eventually, the sixth column indicates the resource groups involved per activity.

Table 4. Total cost of the cataloguing process.

	Cataloguing Process					
	Activity	Average Time (min)	Cost per min (€/min)	Cost (€)	Condition	Resources
Standard Activities	Searching the item on the LMS	0.95	0.94	0.89		Cataloguer + LMS + CM + GO
	New holding description	0.54	0.94	0.50		Cataloguer + LMS + CM + GO
	New item description	0.25	0.94	0.24		Cataloguer + LMS + CM + GO
	Printing the label	0.50	0.94	0.47		Cataloguer + LMS + CM + GO
	Sticking the label on the item	0.47	0.72	0.34		Cataloguer + GO
	Bringing the item to front desk	0.50	0.72	0.36		Cataloguer + GO
	Tagging the item	0.67	1.41	0.94		Cataloguer + LMS + CM + GO + RM + Tag
	Shelving the item	2.80	0.72	2.02		Cataloguer + GO
	Subtotal	6.68		5.77		
Optional Activities	New bibliographic description	5.10	0.94	4.80	*if original_cataloguing*	Cataloguer + LMS + CM + GO
	Mod. bibliog. description	1.24	0.94	1.17	*if copy_cataloguing*	Cataloguer + LMS + CM + GO
	Subtotal	---		---		
	Total Original Cataloguing:	11.78		10.57		
	Total Copy Cataloguing	7.92		6.94		

Note. – LMS = Library Management System. GO = General Overhead. CM = Computer Maintenance. RM = RFID Maintenance.

The subtotal of "standard activities" is the sum of costs included in both processes: original and copy cataloguing. The average time in minutes for standard activities was 6.68 with an activity cost of €5.77. To calculate the total cost of the original cataloguing process, standard activities were summed up with the optional activity, new bibliographic description. The resulting time in minutes was 11.78 with a cost of €10.57. On the other hand, to calculate the total cost of the copy cataloguing process, standard activities were summed up with the optional activity, modify bibliographic description. The results show that the average time for copy cataloguing was 7.92 minutes and its cost was €6.94 per title.

4. ORIGINAL AND COPY CATALOGUING

Because of budgetary constraints and technological changes, cataloguing units have significantly influenced the nature of cataloguing work (**Mitchell, Thompson, & Wu, 2010**). In order to become more efficient, cataloguers constantly search for new ways to increase bibliographic access without spending more money (**Morris & Wool, 1999**). That includes automation, outsourcing, lowered costs for traditional cataloguing, and an increasing variety of information resources to control. With regard to lowering costs, a number of studies on cataloguing costs have been reported in the literature. For instance, the Iowa State University (ISU) conducted a longitudinal time-cost study to investigate the impact of automation on cataloguing costs (**Morris, Hobert, Osmus, & Wool, 2000**). Results showed that in 1987–1990, the average cost of original cataloguing at ISU was $34.13 and that copy cataloguing was $8.18 (**Morris, 1992**). In 1997–1998, original cataloguing costs performed by faculty cataloguers increased to $75.43 and to $58.72 when some original records were contributed by library assistants. Copy cataloguing costs, in the same period, increased to $35.82 when performed by faculty cataloguers and to $8.87 when performed by library assistants. Although costs of original and copy cataloguing increased during the time, **Morris et al. (2000)** also highlighted the fact that the average cost of cataloguing processes (i.e. copy, full original, minimal original and recataloguing) declined from $20.83 to $16.25 between 1990/91 and 1997/98. Authors attributed this decrease to national collaborative efforts, technological development, and reengineering efforts that have improved costs effectiveness and quality process (**Morris & Wool, 1999**). The University of Oregon, following the previous research at ISU, conducted a benchmark analysis during autumn 1997 to determine time and costs for acquisitions, cataloguing, and processing functions (**Slight–Gibney, 1999**). Results showed that the average cost for copy cataloguing was $9.23 per title, and that original cataloguing was $24.92. A third case study by **Ellis-Newman and Robinson (1998)** in an Australian academic library, reported the cost of library services utilizing ABC models. Results showed that the average cost of copy cataloguing at Edith Cowan University was $12.48 and original cataloguing was $54.39. Authors highlighted the importance of implementing activity-based costing systems in libraries to assign more accurate costs to services, to categorize costs, and to develop a price schedule for fee-based services.

In Table 5, all cases cataloguing costs vary library to library, even if the change in the rate of exchange and inflation are included in the three literature case studies. **Charbonneau (2005)** points out that these numbers logically vary because they are based on locally produced data and operations, individual cataloguing expertise, the type of material catalogued, and the cataloguing tools and resources available. Differences in overhead costs can also explain some of the cost variations. For instance, cataloguing overhead costs at ISU represent approximately 45% of the full costs, while in our case study about 10%. Unfortunately, not all studies provide detailed information on the overhead costs. In this case study, these costs include management, secretary, accounting, training, staff meetings and stationery material; however, other university costs such as electricity, telephone, heating and transportation are not accounted for since they are not charged to the library. At ISU overhead costs are organized into overhead centres: paid leave, automation and support services (**Morris et al., 2000**). Some other studies consider overhead costs as approximate since they are estimates or based on incomplete data (**Slight–Gibney, 1999**). Additional factors that vary among libraries are the library structure and process flow, for instance, pre-cataloguing activities are included in the cataloguing costs of one library (e.g. ISU), by the acquisitions department at another library, or even delegated to library assistants or students. In our case study, this activity is not included in the cataloguing flow because it is performed by the person responsible for acquisitions. Finally, differences in the periods of analysis can also explain such cataloguing costs variations, especially due to evolving factors that need to be taken into account. These factors include the increasing automation of cataloguing activities, use of shared cataloguing and authority records, decreasing staffing for cataloguing, and the growing presence of new information formats (**Morris & Wool, 1999**).

Although cataloguing costs data are not necessarily comparable among libraries (**McCain & Shorten, 2002**); "best practices" can be adopted in other libraries to influence their own workflows and to streamline their processes. This case study contributes to the cataloguing cost study literature by providing an additional approach for calculating cataloguing costs based on a fast and simple method as is TDABC. In this case study, the copy cataloguing, as shown in Figure 3, is approximately 34% less costly and 33% less time-consuming than original cataloguing. Thanks to the TDABC's ability to disaggregate costs per activity, it is possible to clearly analyse which activities demand more time, and thus lead to higher costs. For instance, the unique difference between original and copy cataloguing is the creation or modification of a bibliographic description. Results shown in Figure 4 indicate that adapting a bibliographic description from a pre-existing record is approximately 75% less costly than creating a new full bibliographic record. Based on these findings, it is concluded that is worthwhile to recommend librarians to adapt a pre-existing record instead of creating a new bibliographic record from scratch, as this will lead to significant costs and time reductions. Copy cataloguing not only increases cataloguers efficiency by eliminating duplication of effort, but also by reducing typographical errors caused in local libraries (**Beall & Kafadar, 2004**).

Moreover, time equations show that in this particular case, labelling activities such as printing and sticking labels, bringing an item to the front desk, and tagging the item, consume approximately 20–30% of the cataloguing processes. Reviewing the consumed resources revealed that the cataloguer is typically in charge of labelling and shelving catalogued items. A "what-if analysis" can be performed to simulate the effect of delegating these activities to student library employees (SLE). If we take as reference the SLE costs calculated by **Siguenza-Guzman et al. (2014)** of 0.23 €/min, then the resulting costs would be €8.63 for original cataloguing, and €5.00 for copy cataloguing, as shown in Table 6. That is a cost reduction of about 18% for original cataloguing and 28% for copy cataloguing.

Table 5. Cataloguing costs adjusted using inflation and exchange-rates.

Cataloguing costs in the four cases studies

University	Period	Original Cataloguing	Copy Cataloguing	Inflation	Inflated Original Cataloguing	Inflated Copy Cataloguing	EUR Conversion	Original Cataloguing (€)	Copy Cataloguing (€)
Iowa State University (US)	1987–1990	USD $34.13	USD $8.18	68.66%	57.56	13.80	0.72	41.39	9.92
Iowa State University (US)*	1997–1998	USD $75.43	USD $35.82	37.69%	103.86	49.32	0.72	74.68	35.46
Iowa State University (US)†	1997–1998	USD $58.72	USD $ 8.87	37.69%	80.85	12.21	0.72	58.13	8.78
University of Oregon (US)	1997	USD $24.92	USD $ 9.23	39.91%	34.87	12.91	0.72	25.07	9.29
Edith Cowan University (AU)	1998	AUD $54.39	AUD $12.48	47.20%	80.06	18.37	0.74	59.41	13.63
KU Leuven (BE)	2010–2011	EUR €10.57	EUR €6.94	-	-	-	-	10.46	6.83

Note.– Inflation data from inflationdata.com (US, Dec 1987/1998 to Dec 2011), and rba.gov.au (AU, 1998 to 2011).

Exchange data from x-rates.com (USD and AUD to EUR in 2011).

*Cataloguing costs when performed by a faculty cataloguer.

† Cataloguing costs when performed by a library assistant.

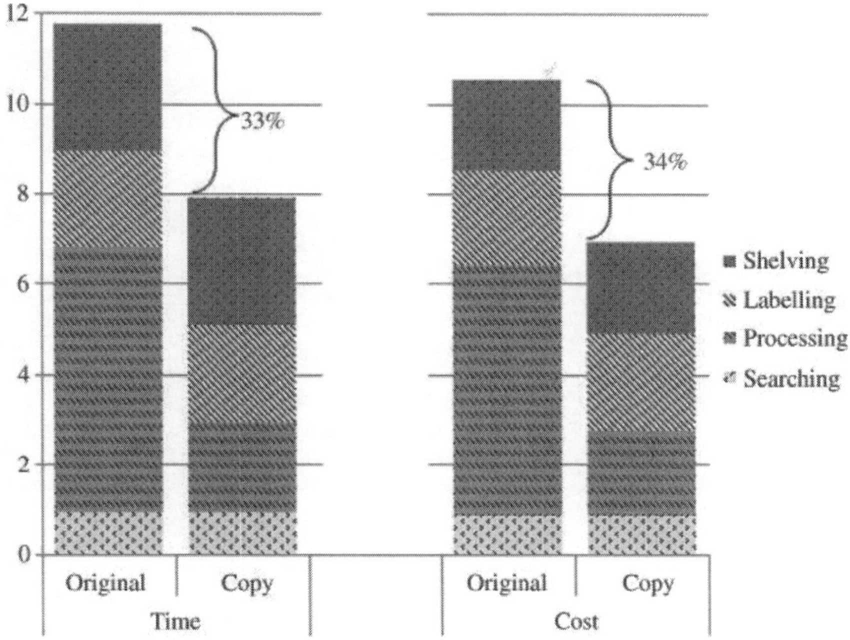

Figure 3. Comparison of original and copy cataloguing in terms of time and cost.

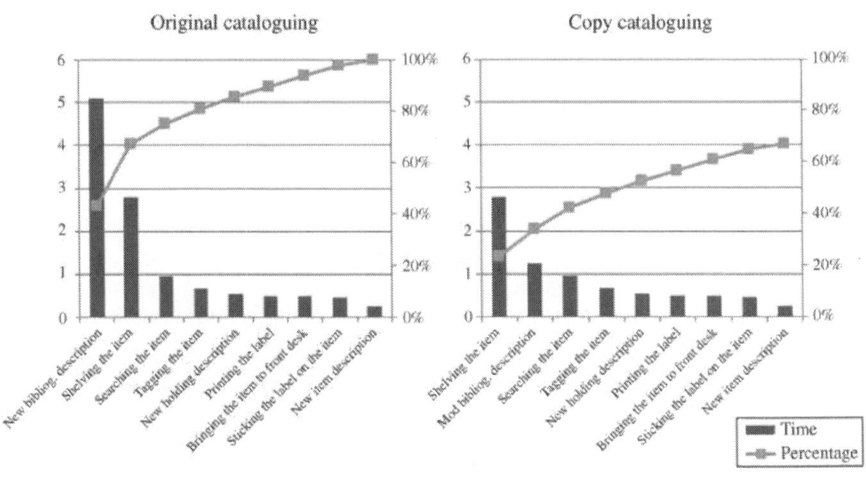

Figure 4. Pareto chart of disaggregating costs per activity for original and copy cataloguing.

Table 6. Example of a what-if analysis applied to the original model of the cataloguing process.

Cataloguing Process						
	Activity	**Average Time (min)**	**Cost per min (€/ min)**	**Cost (€)**	**Condition**	**Resources**
Standard Activities	Searching the item on the LMS	0.95	0.94	0.89		Cataloguer + LMS + CM + GO
	New holding description	0.54	0.94	0.50		Cataloguer + LMS + CM + GO
	New item description	0.25	0.94	0.24		Cataloguer + LMS + CM + GO
	Printing the label	0.50	0.55	0.28		SLE + LMS + CM + GO
	Sticking the label on the item	0.47	0.33	0.15		SLE + GO
	Bringing the item to front desk	0.50	0.33	0.17		SLE + GO
	Tagging the item	0.67	1.02	0.68		SLE + LMS + CM + GO + RM + Tag
	Shelving the item	2.80	0.33	0.92		SLE + GO
	Subtotal	**6.68**		**3.83**		
Optional Activities	New bibliographic description	5.10	0.94	4.80	*if original_ cataloguing*	Cataloguer + LMS + CM + GO
	Mod. bibliog. description	1.24	0.94	1.17	*if copy_ cataloguing*	Cataloguer + LMS + CM + GO
	Subtotal	---		---		
	Total Original Cataloguing:	**11.78**		**8.63**		
	Total Copy Cataloguing	**7.92**		**5.00**		

Note. – LMS = Library Management System. SLE = Student Library Employees. GO = General Overhead. CM = Computer Maintenance. RM = RFID Maintenance.

An additional improvement to the process can be to incorporate batch cataloguing practices and mass retrieval rather than cataloguing individual items or records (**Mitchell et al., 2010**). The process flow analysis enables library managers to group and improve certain activities such as searching, processing (bibliographic, holding and item descriptions), labelling and shelving. These batch activities, such as processing a bunch of items or records at one time, can be performed based on the staff availability at certain moments of the day or week. An example of these sorts of improvements on the cataloguing process is shown in Table 7. In this case, searching, labelling, and shelving are delegated to SLEs and calculated in a batch of 10 items. The duration time of searching items needs to be augmented since SLEs have no expertise in the use of LMS whereas a cataloguer does have. The activity of bringing the item to the front desk is eliminated, suggesting the purchase of an extra rewriting RFID machine be located in the cataloguing computer. Finally, shelving activities were recalculated based on the batching times used by **Siguenza-Guzman et al. (2014)** to model a returning process flow. That includes, an SLE sorting the items in the cluster (i.e. book collection divisions); and then reshelving the items. The resulting time and cost after these improvements can be seen in Table 7.

Results of the what-if analysis indicate that by doing these changes, copy cataloguing is approximately 49% less costly and 42% less time-consuming than original cataloguing. In addition, for original cataloguing, the obtained average time is reduced to 9.26 minutes and the cost to €7.49. This represents about 21% less-time consuming and 29% less-costly than in the real case. Most strikingly, is the case of copy cataloguing in which the reduction is approximately 32% in time and 44% in cost. Therefore, the results of the what-if analysis confirm the validity of implementing these changes in the cataloguing process flow as part of a best practices approach. The enhanced data flow diagram of the cataloguing process is shown in Figure 5.

Table 7. Example of a what-if analysis to improve applied to the original model of the cataloguing process.

	Improvements on the Cataloguing Process					
	Activity	**Average Time (min)**	**Cost per min (€/min)**	**Cost (€)**	**Condition**	**Resources**
Searching	Searching the item on the LMS	1.21	0.55	0.67		SLE + LMS + CM + GO
Processing	Mod. bibliog. description	1.24	0.94	1.17	*if original_cataloguing*	Cataloguer + LMS + CM + GO
	New bibliographic description	5.10	0.94	4.80	*if copy_cataloguing*	Cataloguer + LMS + CM + GO
	New holding description	0.54	0.94	0.50		Cataloguer + LMS + CM + GO
	New item description	0.25	0.94	0.24		Cataloguer + LMS + CM + GO
Labelling	Printing the label	0.50	0.55	0.28		SLE + LMS + CM + GO
	Sticking the label on the item	0.47	0.33	0.15		SLE + GO
	Tagging the item	0.67	1.02	0.68		SLE + LMS + CM + GO + RM + Tag
Shelving'	Classifying the item	0.17	0.33	0.06		SLE + GO
	Shelving the item	0.35	0.33	0.12		SLE + GO
	Total Original Cataloguing:	**9.26**		**7.49**		
	Total Copy Cataloguing	**5.40**		**3.85**		

Note. – LMS = Library Management System. SLE = Student Library Employees. GO = General Overhead. CM = Computer Maintenance. RM = RFID Maintenance.

*Shelving time taken from **Siguenza-Guzman et al. (2014)**.

Modified Cataloguing Process

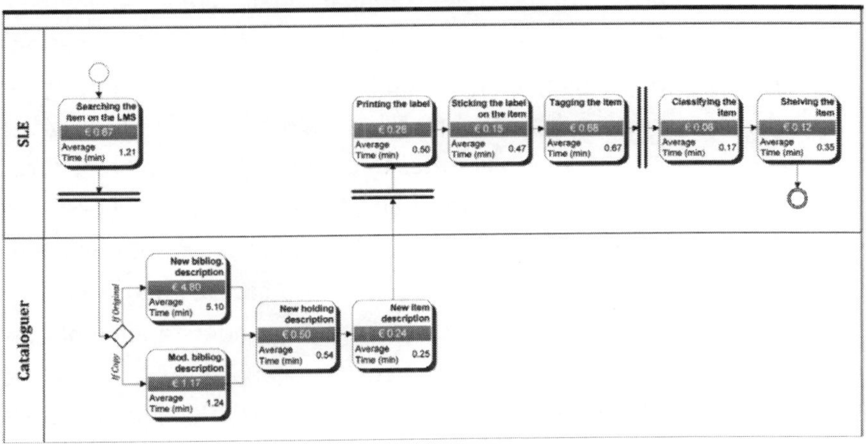

Note. – LMS = Library Management System.

Figure 5. Improvements on the cataloguing process.

5. BENEFITS OF TDABC IN CATALOGUING PROCESSES

The implementation of TDABC in cataloguing processes at CBA demonstrated important benefits. The first important benefit, shows the possibility to clearly*discriminate activities regarding the time*, and thus determining which activities demand more time and cost. For instance, analysis showed that labelling and shelving activities consume unnecessary resources. The second benefit is a consequence of the previous one. That is, the possibility of performing "what-if analysis" to *simulate potential scenarios;* for example, as labelling and shelving consumed unnecessary resources, what-if analyses were conducted. The first simulation showed that by delegating these activities to SLE staff, the library manager could reduce up to 23% of costs. The second simulation improves the previous simulation by incorporating batching processes. Results showed a reduction of approximately 37% of cataloguing costs, allowing librarians to improve their processes and liberate their time to perform other specialized activities. An additional benefit is that TDABC allows *benchmarking different scenarios locally and among libraries*, for example, original versus copy cataloguing. These results show that original cataloguing in this particular scenario is 30% more time-consuming, and consequently more costly than original cataloguing. Moreover, by benchmarking time and cost of activities among libraries, TDABC allows libraries to adopt policies and procedures to improve efficiency. When sharing benchmark figures, Slight-Gibney (1999) states that for other libraries, the time spent on various activities is probably more useful than the costs. In the case of benchmarking cost, the participating libraries require a common understanding of how to attribute indirect costs to their calculations and of what costs are included/excluded in order to have standard results. This problem does not occur when benchmarking the time spent on different activities. Eventually, a fourth important benefit is that TDABC allows *justifying decisions and choices.*TDABC allows both managers and staff to better understand alternative options and accept the need for change; for example, the decision of transferring responsibilities from cataloguers to SLE staff, as well as, structuring batch activities (processing, labelling and shelving). In fact, library managers should constantly analyse their cost information and keep their models updated in order to redesign workflows efficiently and effectively, as well as to reallocate resources and tasks.

Nevertheless, **Siguenza-Guzman et al. (2013b**, **2014**) suggest important considerations to be borne in mind at implementing TDABC in academic libraries, namely: 1) the resource intensity of data collection to gather the time duration of activities, as well as to document the activity flows; 2) the need of a dedicated software tool to keep the flows updated and consequently to facilitate long-term maintenance; and 3) the commitment of library managers and staff during the data collection.

6. CONCLUSION

In this paper, the TDABC implementation was described in two main cataloguing activities, namely, original and copy cataloguing. These processes were selected because they are considered to be a part of the core activities of a library to manage its collection, but are also resource intensive. Based on our findings, we can conclude that TDABC is a quicker and easier way of calculating cataloguing savings. In fact, TDABC is a useful method to perform cost analysis in cataloguing processes, and consequently provides valuable data for managerial decisions. The TDABC implementation provided library managers with important information about cataloguing costs and performance measurements; and guided decisions concerning resource allocation and process improvements. For example, based on the obtained results, the library manager decided to delegate certain activities and define a set of batch activities. This case study, therefore, shows significant contributions to the literature on the implementation of advanced cost models for library processes, and more precisely for cataloguing activities.

A potential direction for future research is to expand this study to different cataloguing activities such as cooperative, contract, and outsourcing cataloguing. TDABC can also offer the possibility to discuss how these trends in cataloguing processes affect cataloguing units. TDABC will allow to examine whether these trends really provide an opportunity for cataloguers to spend their time on other cataloguing activities, such as enhancing existing records, or not. Another interesting area for future research would be a similar analysis for cataloguing audio-visual items and other special materials such as old books, non-print materials, and maps. Utilizing TDABC to benchmark libraries for "best practices" is an additional prospect for future analysis.

ACKNOWLEDGMENTS

We wish to thank the staff of the Arenberg Campus Library at KU Leuven for their cooperation throughout this study; Hilde Van Kiel, head of the Library, for her big support and guidance; and Marc Verbrugge for his assistance during the data collection phase. The authors also want to thank Paul Vanegas for his comments on the paper. This research project was funded by the University of Cuenca, the Flemish Interuniversity Council (VLIR-IUC), and the National Secretariat of Higher Education, Science, Technology and Innovation of Ecuador (SENESCYT).

NOTES

[1]http://bib.kuleuven.be/cba/

REFERENCES

1. ACRL Research Planning and Review Committee. (2010). 2010 top ten trends in academic libraries: A review of the current literature. College & Research Libraries News, 71(6), 286–292
2. Beall, J., & Kafadar, K. (2004). The effectiveness of copy cataloging at eliminating typographical errors in shared bibliographic records. Library Resources & Technical Services, 48(2), 92–101
3. Campus Bibliotheek Arenberg. (2013, December 23). CBA - Campusbibliotheek Arenberg. Retrieved January 22, 2014, from https://bib.kuleuven.be/cba/cba-english/home
4. Charbonneau, M. D. (2005). Production benchmarks for catalogers in academic libraries. Library Resources & Technical Services, 49(1), 40–48
5. Ching, S. H., Leung, M. W., Fidow, M., & Huang, K. L. (2008). Allocating costs in the business operation of library consortium: The

case study of Super e-Book Consortium. Library Collections, Acquisitions, and Technical Services, 32(2), 97–103.

6. Cooper, R., & Kaplan, R. S. (1988). Measure costs right: Make the right decision. Harvard Business Review, 66(5), 96–103
7. Cottrell, T. (Terry). (2012). Three phantom budget cuts and how to avoid them. Bottom Line: Managing Library Finances, The, 25(1), 16–20.
8. Ellis-Newman, J. (2003). Activity-Based Costing in user services of an academic library. Library Trends, 51(3), 333–348
9. Ellis-Newman, J., Izan, H., & Robinson, P. (1996). Costing support services in universities: an application of activity-based costing. Journal of Institutional Research in Australasia, 5(1), 75–86
10. Ellis-Newman, J., & Robinson, P. (1998). The cost of library services: Activity-based costing in an Australian academic library. The Journal of Academic Librarianship, 24(5), 373–379.
11. Everaert, P., Bruggeman, W., Sarens, G., Anderson, S. R., & Levant, Y. (2008). Cost modeling in logistics using time-driven ABC: Experiences from a wholesaler. International Journal of Physical Distribution & Logistics Management, 38(3), 172–191.
12. Goddard, A., & Ooi, K. (1998). Activity-Based Costing and central overhead cost allocation in universities: A case study. Public Money and Management, 18(3), 31–38.
13. Hoozée, S., Vermeire, L., & Bruggeman, W. (2012). The impact of refinement on the accuracy of Time-Driven ABC. Abacus, 48(4), 439–472.
14. Kaplan, R. S., & Anderson, S. R. (2004). Time-Driven Activity-Based Costing-Tool Kit. Harvard Business Review, (82), 131–138
15. Kaplan, R. S., & Anderson, S. R. (2007). Time-Driven Activity-Based Costing: A simpler and more powerful path to higher profits. Boston, MA, USA: Harvard Business School Press

16. Kaplan, R. S., & Cooper, R. (1998). Cost & Effect: Using integrated cost systems to drive profitability and performance. Boston, MA, USA: Harvard Business School Press

17. Kont, K.-R. (2011). New cost accounting models in measuring of library employees' performance. Library Management, 33(1/2), 50–65.

18. Manaf, Z. A., & Rahman, R. A. (2006). Examining the quality of National Library of Malaysia (NLM) cataloguing in publication (CIP) records. Library Review, 55(6), 363–373.

19. McCain, C., & Shorten, J. (2002). Cataloging efficiency and effectiveness. Library Resources & Technical Services, 46(1), 23–31.

20. Mitchell, A. M., Thompson, J. M., & Wu, A. (2010). Agile cataloging: Staffing and skills for a bibliographic future. Cataloging & Classification Quarterly, 48(6-7), 506–524.

21. Morris, D. E. (1992). Staff time and costs for cataloging. Library Resources & Technical Services, 36(1), 79–95

22. Morris, D. E., Hobert, C. B., Osmus, L., & Wool, G. (2000). Cataloging staff costs revisited. Library Resources & Technical Services, 44(2), 70–83

23. Morris, D. E., & Wool, G. (1999). Cataloging: Librarianship's best bargain. Library Journal, 124(11), 44–46

24. Novak, D. D., Paulos, A., & Clair, G. S. (2011). Data-driven budget reductions: A case study. The Bottom Line: Managing Library Finances, 24(1), 24–34.

25. Pernot, E., Roodhooft, F., & Van den Abbeele, A. (2007). Time-Driven Activity-Based Costing for inter-library services: A case study in a university. The Journal of Academic Librarianship, 33(5), 551–560.

26. Reitz, J. M. (2004). ABC-CLIO - ODLIS. ODLIS - Online dictionary for library and information science. Libraries Unlimited. Retrieved December 18, 2013, from http://www.abc-clio.com/ odlis/ odlis_a.aspx

27. Roberts, S. A. (2003). Financial management of libraries: Past trends and future prospects. Library Trends, 51(3), 462–493

28. Siguenza-Guzman, L., Van den Abbeele, A., Vandewalle, J., Verhaaren, H., & Cattrysse, D. (2013a). Recent evolutions in costing systems: A literature review of Time-Driven Activity-Based Costing. ReBEL - Review of Business and Economic Literature, 58(1), 34–64

29. Siguenza-Guzman, L., Van den Abbeele, A., Vandewalle, J., Verhaaren, H., & Cattrysse, D. (2013b). Towards a holistic analysis tool to support decision-making in libraries. In Proceedings of the IATUL Conferences. Cape Town, South Africa: Purdue e-Pubs. Retrieved December 18, 2013, from http://docs.lib.purdue.edu/iatul/2013/papers/29/

30. Siguenza-Guzman, L., Van den Abbeele, A., Vandewalle, J., Verhaaren, H., & Cattrysse, D. (2014). Using Time-Driven Activity-Based Costing to support library management decisions: A case study for lending and returning processes. Library Quarterly: Information, Community, Policy, 84(1), 76–98.

31. Slight–Gibney, N. (1999). How far have we come? Benchmarking time and costs for monograph purchasing. Library Collections, Acquisitions, and Technical Services, 23(1), 47–59.

32. Stouthuysen, K., Swiggers, M., Reheul, A.-M., & Roodhooft, F. (2010). Time-Driven Activity-Based Costing for a library acquisition process: A case study in a Belgian University. Library Collections, Acquisitions, and Technical Services, 34(2–3), 83–91.

33. Vazakidis, A., & Karagiannis, I. (2009). Activity-Based Management and traditional costing in tourist enterprises (A hotel implementation model). Operational Research, 11(2), 123–147.

34. Wegmann, G., & Nozile, S. (2009). The Activity-Based Costing method developments: State-of-the art and case study. The IUP Journal of Accounting Research and Audit Practices, 8(1), 7–22.

35. Yilmaz, R. (2008). Creating the profit focused organization using Time-Driven Activity Based Costing. In EABR & TLC Conferences Proceedings (p. 8). Salzburg, Austria: Clute Institute for Academic Research

Chapter 9

Scholarly social media profiles and libraries: A review

Judit Ward[1], William Bejarano[1], Anikó Dudás[2]

[1]Rutgers, The State University of New Jersey, USA

[2]Pázmány Péter University, Budapest, Hungary

ABSTRACT

This article aims to point out emerging roles and responsibilities for academic librarians with the potential of better integrating the library in the research process. In order to find out how to enhance the online reputation and discoverability of individual faculty members as well as their affiliated institutions, the authors worked side-by-side with researchers in the United States and Europe to explore, create, revise, and disambiguate scholarly profiles in various software applications. In an attempt to understand and organize scholarly social media, including new, alternative metrics, the authors reviewed and classified the major academic profile platforms, highlighting the overlapping elements, benefits, and drawbacks inherent in each. The consensus is that it would be time-consuming to keep one's profile current and accurate on all of these platforms, given the plethora of underlying problems, also discussed in detail in the article. However, it came as a startling discovery that reluctance to engage with scholarly social media may cause a misrepresentation of a researcher's academic achievements and may come with unforeseen consequences. The authors claim that current skills and competencies can secure an essential role for academic librarians in the research workflow by means of monitoring and navigating researcher profiles in scholarly social media in order to best represent the scholarship of their host institutions.

KEYWORDS

altmetric; discoverability; researcher profile; scholarly selfie; scholarly social media

1. INTRODUCTION

In addition to subject-specific reference questions and literature searches, academic librarians all over the world have been assisting users routinely with topics related to citation analysis and bibliometrics in support of their research, scholarly publication, and communication. The queries are mostly related to traditional metrics, including journal impact factors and individual performance indicators, such as citation count or h-index.

More recently, however, requests also indicate a strong interest and diverse information need regarding alternative metrics, such as individual article download counts or some particular metrics provided by one of the numerous popular scholarly social media applications. While trying to make sense of the copious metrics offered by these evolving new platforms, the common feature of these inquiries led to two major questions. First, *how is it possible to compare scholarly output across these platforms*? For example, since the citation count of the same author in Web of Science, in conjunction with all metrics derived from it, significantly differs from the number provided by Google Scholar Citations, which one is correct? The second, closely related question, *where should one create an author profile?*, expresses reasonable concerns about the longevity of scholarly social media applications.

In order to answer these two seemingly simple questions, one should be familiar with not only the research workflow and the evaluation of scholarly performance, but also the current social media applications. Thus, added here is a third question, compelling for the librarian community: *what can librarians do to help researchers navigate the maze of these applications?*

2. SOCIAL MEDIA IN ACADEMIA

Researchers are known to follow everyday social media, such as Facebook and Twitter, with great reluctance (**Mas-Bleda, Thelwall, Kousha, & Aguillo, 2014**;**Mendez et al., 2009**). However, ignoring the scholarly applications may lead to a misrepresentation of their academic achievements and may come with unforeseen consequences, not only to them and their institutions, but also to their colleagues and students (**Hajnal Ward, Bejarano, & Dudas, 2014**). The complete lack of an online presence can limit a researcher's visibility, and undesirable search results may be discovered before desirable ones (**Bik & Goldstein, 2013**).

Scholarly reputation and discoverability of research have always been a top priority for scientists (**Hoover, 2014**; **Peters, 2014**; **Woelfle, Olliaro, & Todd, 2011**). With the increasing popularity of collaborative and citation management applications, as well as new metrics to track scholarly impact, social media gained a new importance in scholarly communication. Peer use and word of mouth quickly resulted in the acceptance of one or more scholarly social media platforms within a particular academic community. According to a survey conducted by *Nature* in May 2014, based on the data provided by about 3,500 scientists from 95 countries, Google Scholar is known by over 70 per cent of the respondents. ResearchGate has been found to be the best known scholarly social media platform (better known among scientists and engineers than the popular social media outlet Twitter) with 29 per cent registered in the past year. These figures refer only to scholars' familiarity with the applications, not their satisfaction. The survey also revealed that the majority of the respondents simply created and maintain their profile without using other, more advanced features offered by the applications (**Van Norden, 2014**). It has been also claimed that "a growing body of evidence suggests that public visibility and constructive conversation on social media networks can be beneficial for scientists, impacting research in a number of key ways" (**Bik & Goldstein, 2013**).

Social media for scientists have recently become a hot topic. For all the reluctance on behalf of scientists, networks proliferate (**Mangan, 2012**), tools have a potential to measure research impact (**Haustein & Siebenlist, 2011**), and scholarly social media have been reviewed extensively (**Bik & Goldstein, 2013;Ortega, 2015**). Different fields of science are reported to use social media in academia such as biology (**Crawford, 2011**), technology, engineering, and mathematics (**Galloway, Pease, & Rauh, 2013**; **Giglia, 2011**), and humanities (**Hammarfelt, 2014**). Researchers are inundated with introductions as well as detailed guidelines about how to use social media in general (e.g., **Goodier & Czerniewicz, 2012**; **Gruzd, 2012**), how to benefit the most from engaging in social media (**Bik & Goldstein, 2013**), and how social media can enhance their visibility (**Bar-Ilan et al., 2012**; **Giglia, 2011**; **Goodier & Czerniewicz, 2012**).

Well-researched are the various methods of constructing metrics from social media for academic purposes, such as counting twitter citations to estimate the likely impact of articles (**Eysenbach, 2011**). The recently launched and still emerging field of altmetrics (**Priem, Groth, & Taraborelli, 2012a**) suggests new indicators of scholarly impact. The literature tends to be concerned with impact measures rather than scholarly communication itself or scholarly communication networks (**Thelwall, Haustein, Larivière, & Sugimoto, 2013**). Several studies analyse the relationship between traditional and alternative measures of scholarly reputation, (**Ortega, 2015**; **Torres-Salinas, Cabezas-Clavijo, & Jiménez-Contreras, 2013**; **Weller & Strohmaier, 2014**), and the implications of altmetrics on academic and research libraries (**Rodgers & Barbrow, 2013**). In contrast, there is surprisingly little research about the scholarly social network platforms themselves, and there is even less literature on researcher profiles and the impact of these profiles on scholarly communication, collaboration, and research.

"Profile" is listed first among the eight distinctive functions of social media for scholarly use in a recent summary of the academic use of Social Network Sites (SNS) (**Nentwich & König, 2014**). *Userprofiles* are "digital representations of users and as such, the central nodes of SNS. Various kinds of information can be made available to other members in a pre-structured way, from contact information to tracking of user activities. In some SNS it is also possible to have specific profiles for organizations. Thus, profiles are like enhanced calling cards of individuals, organizations, and groups. Some SNS experiment with special scores to automatically rate user activity on the basis of their activity in the SNS, thereby creating a potential metric for reputation (e.g. RG Score in *ResearchGate*)" (**Nentwich & König, 2014**, pp. 110–111).

The big surprise of the past few years has been the sudden and unexpected popularity of the new scholarly profiles, created and/or regulated by scholars, who, perhaps not too long ago, were unwilling to embrace social media. The researcher profile is not merely an ephemeral snapshot at a specific point in time, but instead it serves as a scholarly portrait of a scientist, and has a great potential to promote the scholar's oeuvre, discoverability, reputation, and, in the long run, citations. Scholarly social media platforms generally interpret science in the broadest sense and encourage open access to not only scholarly articles, but presentations, data sets, negative results, grey literature, notes, drafts, and so on (**Piwowar, 2013**; **Priem, Taraborelli, Groth, & Neylon, 2010**; **Thelwall & Kousha, 2014**; **Weller, 2015**). Often granting access to full-text articles, the "scholarly selfie"1 also provides more information on the scholar's research outside the traditional publishing spheres. The new indicators used in the researcher profiles complement the traditional metrics, and have the potential to provide a more complete view of the scholar's impact (**Fenner, 2014**; **Ortega, 2015**).

Researcher profiles in the scholarly social media follow the general practice of social networking applications. Scholars have to register, create, and/or claim their profile by setting up an account, which often requires affiliation to an educational or research institution. The platform serves as a collaboration site with plenty of opportunities to share their research, follow other scholars' work, and offer and solicit opinions on work in progress. Based on the articles and other research-related documents scholars upload to these platforms, statistical data can be monitored and collected on traditional and new metrics, including views, download counts, shares, or citations. The scholarly social media also became a venue to find research partners, co-authors, and collaborators, as well as new jobs (**Hajnal Ward et al., 2014**). The structure of the scholarly social network "could be a valuable data source to explore aspects of informal scholarly communication, especially if the new academic social websites revolutionize research as much as the Internet did" (**Thelwall & Kousha, 2014**, p. 729).

3. SCHOLARLY PERFORMANCE INDICATORS IN ACADEMIC LIBRARIES

Librarians have assumed leading roles in the advancement of new technologies and pioneered meaningful applications related to bibliometrics and social media. Academic librarians keep abreast of new advancements and technologies related to research, while researchers and librarians have both developed a keen interest in figures to quantify research. Several bibliometrics topics were covered by librarians and information specialists (**Bar-Ilan et al., 2012**; **Cronin, 2001**; **De Groote, 2008**; **Hood & Wilson, 2001**; **Jacso, 2008**; **Leydesdorff, 1998**), and scholars from a variety of fields have also published extensively on topics related to the issues in bibliometrics (**Abt, 2000**; **Boyack, Klavans, & Börner, 2005**; **Hall, 2014**; **Hicks, 2005**; **Hirsch, 2005**; **Schreiber, 2008**; **Vinkler, 2010**). Scientists and librarians working in conjunction on citation analysis agree that the output indicators are just as good as the data input (**Cronin, 2001**; **De Bellis, 2009**;**Jacso, 2008**; **Meho & Yang, 2007**; **Noruzi, 2005**; **West & McIlwaine, 2002**). The situation is no different in regard to scholarly social media.

3.1. Traditional metrics

Scholarly impact is traditionally assessed by statistics and complex mathematical formulas, which quantify research output and scholarly reputation in various contexts. The science of bibliometrics has been around for over fifty years, providing tools for the quantitative analysis of academic publications (**Bornmann, Mutz, Neuhaus, & Daniel, 2008**; **Boyack et al., 2005**; **Meho, 2007**;**Moed, 2006**; **Vinkler, 2010**). In the long run, as they represent scholarly impact, they are considered as established factors, which are used in making hiring, promotion, and funding decisions, among others.

Bibliometric indicators may represent the reputation of an individual scholar, such as total publication count and citation count (total and average); a journal (the impact factor from Thomson Reuters Scientific being the most accepted); an institution; or even an entire country. In addition to the most commonly used indicators for authors, new metrics, such as the h-index (**Hirsch, 2005**), the g-index (**Egghe, 2006**), and further variations are also used to measure the researcher's scholarly output (**Schreiber, 2008**). Similarly, in the case of journals, in addition to the well-known impact factor, new metrics have evolved, such as the Eigenfactor or SJR (SCImago Journal Rank) based on the Google PageRank™ algorithm (**Hajnal Ward, Stewart, Cox, Candon, & Cook, 2011**; **Moed, 2006**; **Vinkler, 2010**).

Sources of traditional bibliometrics include Web of Science (with its databases provided by Thomson Reuters Scientific) and Scopus by Elsevier. Used mostly in natural sciences and social sciences, their proprietary data have been instrumental for decision makers in hiring, promoting, or funding scholars. Their robust data, illustrated with visually pleasing evaluation charts and graphs, are displayed on an individual author's profile. The Web of Science profile tracks the scholar's activities via the total number of publications and citation count, as well as metrics computed based on these figures, such as the h-index or average citation count. Drawn from the entire database, citation maps can show the scholarly impact of any given article by displaying citing articles.

The third important bibliometrics tool, Google Scholar, is open and free to everyone, and as such, is widely used all over the world. Crawling its own collection, Google Scholar calculates similar metrics, such as citation counts. It should be noted that each database computes its metrics from its own data, which can result in inaccuracies and overinflated metrics (**Hajnal Ward et al., 2011**; **Jacso, 2008**; **Meho & Yang, 2007**).

3.2. Alternative metrics

Traditional citation metrics are far from being perfect, and as such, have been criticized extensively (**Abt, 2000**; **Gorraiz, Gumpenberger, & Schlögl, 2014**;**Moed, 2006**; **Sud & Thelwall, 2014**). With a new approach to science (open access, open data, open repositories, open science), new horizons opened up in evaluating scholarly performance with several tools to gather alternative metrics, including PLoS Article-Level Metrics, ReaderMeter, CitedIn, total-impact, altmetric.com, and ScienceCard (**Weller, 2015**). The increasing popularity of social media applications resulted in new ways of scholarly communication and new metrics proliferated to assess scholarly impact called altmetrics (**Cronin, 2001**; **Fenner, 2014**; **Galloway et al., 2013**; **Piwowar, 2013**;**Priem et al., 2012a**; **Priem, Piwowar, & Hemminger, 2012b**; **Sud & Thelwall, 2014**). Altmetrics is defined as “the creation and study of new metrics based on the Social Web for analysing, and informing scholarship,” in the Altmetrics Manifesto by Priem et al. in 2010 on their site (**Priem et al., 2010**). The term is used interchangeably with alternative metrics, or for any kind of non-traditional scholarly performance indicator, while “altmetric” is branded as a subscription-based service at altmetric.com to compile article level metrics by providing “a cluster of servers that watch social media sites, newspapers, government policy documents and other sources for mentions of scholarly articles” (**What does altmetric do?, 2015**).

The web-based alternative indicators of scientific impact show the potential of measuring any kind of scholarly activity. Traditional metrics consider journal publications as the cornerstone of the author’s work, while the newer metrics can be computed on the basis of other research-related data and output, such as data sets, blog posts, or white papers, just to mention a few, as sources. In addition to citations of these items, new metrics can cover any data showing impact such as page views, downloads, comments on the text, mentions in tweets or posts, or social bookmarks (for more, see literature review on altmetric in (**Thelwall et al., 2013**). There seems to be a correlation between traditional and alternative metrics. Recent study results show that the most cited articles within recent years also have significantly higher altmetrics indicators (**Torres-Salinas et al., 2013**).

The new measurements derived from the social web are increasingly advocated and widely used as early indicators of article impact and usefulness (**Thelwall et al., 2013**). Several studies have investigated alternative metrics and their relationship with traditional citation indicators (**Bornmann, 2014**; **Ortega, 2015**;**Torres-Salinas et al., 2013**), and found statistically significant associations between higher metric scores and higher citations for articles in six of the eleven altmetric (Twitter, Facebook wall posts, research highlights, blogs, mainstream media and forums) in medical and biological sciences and for articles with at least one altmetric mention (**Eysenbach, 2011**; **Thelwall et al., 2013**).

Alternative metrics are claimed to provide an opportunity for institutions and researchers to bridge informal academic discourse with the formal output of research. As more scholars move their communication to open social spaces like public blogs, social networks, and Open Access journals, this discourse and its impact becomes traceable and measurable (**Rodgers & Barbrow, 2013**). Social media and article-level metrics may thus be particularly important for unveiling research impacts that cannot be reflected in traditional scientific metrics. For example, some articles may be rarely cited, but heavily read and downloaded by academics (**Priem et al., 2012b**). The social web is also suggested to help satisfy the need for timely metrics. An article may be publicly endorsed within hours of publication, mentioned in a tweet or blog, discussed on a forum or a social network site, and saved in a reference manager. These can be considered as proof of instant impact of the scholar's work (**Sud & Thelwall, 2014**), and actual use from the user's perspective (**Gorraiz et al., 2014**). Another major argument for the use of alternative metrics lies in the different interpretation of impact. Citations only assess the impact of scholarly literature on those who cite, which completely neglects other audiences of scholarly literature who may read, but do not cite (**Thelwall et al., 2013**).

The new metrics assume that scholars are actually engaged in the online social environments, and interact with scholarly products in the social web. Research shows a variety of the use and coverage of social media environments (**Haustein et al., 2014**). Researchers are encouraged in various ways to take advantage of the scholarly social media, and new rules reflect some recognition of unorthodox research output. The National Science Foundation started to request scholarly "products" in the biosketch along with the grant applications instead of publications starting in 2013, which indicates that these new advancements have been in the focus of attention (**Piwowar, 2013**). The new format of the NIH Biosketch, effective after May 25, 2015, also shows changes in the language, using "contributions to science" instead of "peer-reviewed publications" as section heading (**NIH biosketch sample document**). The European research collaboration ACUMEN (Academic Careers Understood through Measurement and Norms) has developed criteria and guidelines for Good Evaluation Practices (GEP), a flexible framework for a standardized individual researcher's portfolio, including multiple webometric and altmetric indicators as well (**ACUMEN Portfolio, 2014**).

A scholar's presence in the social media will probably not replace the reputation computed by traditional bibliographic methods. As one of the latest responses to alternative methods of measuring scholarly impact, a brand new metric has been computed to indicate as a measure of discrepancies between a scientist's social media profile and actual scholarly activities. The "Kardashian Index" plots the total citations of a sample of researchers-twitterers against the number of their Twitter followers only to point out the so-called "scientific Kardashians"2, i.e., the ones whose actual impact, based on scholarly citations, does not measure up with their social media presence (**Hall, 2014**).

3.3. Scholarly selfies

A major development in the assessment of scholarly impact is the sudden expansion of author profiles or researcher profiles, called here "scholarly selfies", both in proprietary databases and in social media. The word "selfie" was voted the word of the year of the Oxford Dictionaries in 2013 with a unanimous decision (**The Oxford Dictionaries..., 2013**). By their definition, a selfie is "a photograph that one has taken of oneself, typically one taken with a smartphone or webcam and shared via social media". Adding the adjective "scholarly" expands this meaning to denoting the snapshot a researcher has taken and shared on a scholarly social media platform at a given time, resulting in a scholarly profile, sometimes including a photo. The concept also indicates that researchers are willing to take control over their online reputation by claiming responsibility over their presence.

The above mentioned alternative metrics are widely used by the scholarly profiles to complement traditional scholarly performance indicators. Simultaneously with the growing popularity of the new metrics, scholarly social media profiles have become commonly accepted in the academic communities, proven by the large number of scientists joining one or more of these applications (**Crawford, 2011**; **Foley & Kochalko, 2012**; **Goodier & Czerniewicz, 2012**; **Gruzd, 2012**; **Hajnal Ward et al., 2014**; **Mangan, 2012**; **Thelwall & Kousha, 2014**).

The idea of a researcher profile is not new, since proprietary databases also feature their authors by using the publication data from their own sources and presenting them as an author page. Web of Science and Scopus both provide a profile page for authors whose articles are indexed in their databases. Traditional author profiles include author's institutional homepages, and profiles mandated or suggested by an institution (such as the Rutgers University Faculty Survey athttp://oirap.rutgers.edu/surveys/facsurv.html), a grant funding organization (such as the NIH Biosketch and author profile tool MyNCBI athttp://www.ncbi.nlm.nih.gov/sites/myncbi), a national academic administration (such as the Hungarian Scientific Works – Magyar Tudományos Művek Tára, MTMT https://www.mtmt.hu), or a country's portal for accredited doctoral schools with the research profiles of the core-member professors and supervisors (such as the Hungarian Doctoral Data Base athttp://www.doktori.hu) (**Hajnal Ward et al., 2014**).

Scholarly selfies, however, follow the rules of social media applications. Researchers have to set up an account, or, in other versions, claim a pre-fabricated one, and build a profile, similar to LinkedIn or Facebook. Then they can read newsfeeds, join groups, make friends, follow other researchers' activities, and use the forum for communication about their research. A scholarly social media platform also features bibliographic records of the researcher's publications, either from its own database or uploaded by the author. The latter allows listing the previously mentioned unorthodox products of research, such as datasets and even negative results. Based on the records in the author's profile, statistical data on views, downloads, and citations are instantly available. Tagging their research interests provides a great opportunity for discoverability and access for potential co-authors. Many platforms offer a Q&A section to discuss scholarly issues or just bounce off ideas as well as job postings (**Bik & Goldstein, 2013**; **Rinaldi, 2014**).

Unbeknownst to the scholar, the most popular applications may generate researcher profiles from their own database or through their established search engines. With their strong focus on particular research areas, scholarly social media applications, such as ResearchGate, Academia.edu, and ORCID, offer platforms to document scholarly output, using both established and unorthodox metrics. The applications promote discoverability and increase opportunities for collaboration. Some can be helpful in the daily work of researchers, such as those that offer file sharing and citation management capabilities, while others have solved the problem of disambiguating authors with their unique researcher IDs (**Bik & Goldstein, 2013**; **Foley & Kochalko, 2012**; **Haak, Fenner, Paglione, Pentz, & Ratner, 2012**; **Hajnal Ward et al., 2014**; **Rinaldi, 2014**; **Thelwall & Kousha, 2014**; **Thelwall & Kousha, 2015**).

4. RESEARCHER PROFILES IN THE SCHOLARLY SOCIAL MEDIA

A classification of social media networks used for scholarly purposes (**Nentwich & König, 2014**) lists eight main functions, with the main topic of the current study, ***user profiles***, as first on the list. The second, ***communication***, includes messaging, chatting, discussion forums/groups, microblogging, nudging, and videoconferencing. The third is ***networking***, same as on Facebook with lists of contacts/friends, searches for more contacts, search functions (both partly automated), invitations, etc. The next function is called "***directing attention***", and relates to features such as displaying current issues on the opening page ("feed"), external notifications (via e-mail or messages), and the "Like" and "Share" buttons. The next function is related to ***groups***: members can find others with similar interests, can benefit from the digital environment for discussion and collaboration with services such as discussion forum, file upload, collaborative writing environments, tools to administer participants in events, selective access to groups, and passive membership. Another function is the built in ***calendar*** to coordinate dates, plan, and market events. A valuable feature is the***literature-related functions*** such as searching for academic - literature by giving access to other databases, publication lists and database entries of members; compiling bibliographies; open access archive; various attention direction services like notifications, based on topicality, semantic relationships, "Have read" buttons, commenting or rating, "Share this" function, access statistics, and visualization of networks of co-authors. Finally, a miscellaneous category called ***further services*** may include specialized and target-group-specific services, for example job exchange services, blogging, embedding of services of external providers via apps, and advertisement (**Nentwich & König, 2014**). Scholarly social media applications use more or less of these functions, depending on the main purpose of the platform.

A user profile, created by the researcher, has a great potential to mirror the author's scholarship, which includes the choice of how much they wish to disclose online about themselves and their work. Five main types of academic persona were observed on the social web with fluid boundaries between them: formal, networked, comprehensive, teaching and uncontainable (**Barbour & Marshall, 2012**). The rather static and simple ***formal self*** has minimal interactivity and resembles earlier generations of Web sites; institutional faculty pages could be the typical example. Close to this in functionality are the MTMT and MyNCBI with SciENcv. Scholarly social media profiles present the ***networked self***, a more public presentation of the individual within the traditional academic frame, focusing on sharing ideas and networking. The professional social media platform LinkedIn can be used for these networking purposes, but tends to be used more in a "formal self" capacity, so it can be classified somewhere in between the two. The online persona called the ***comprehensive self*** uses social media for both academic and personal purposes. In addition to a scholarly online presence, one also keeps in touch with friends and family members via social media applications. Scholars engaged in instructional activities may favour a special ***teaching self*** with the focus on students as opposed to colleagues, a great opportunity to connect with digital native students. Finally, opting out without engaging in social media at all does not mean that one has no online presence. This online persona is labelled as the ***uncontainable self***, which comes with the risk that others will create one for them, and might be criticized or even defamed, for example by students using sites such as ratemyprofessor.com (**Barbour & Marshall, 2012**).

The extent of social media engagement varies among those involved. Five profile types were observed according to activity levels and usage intensities (from lowest to highest): me-too presence, digital calling card, passive networking, active networking and communication, and cyberentrepreneurship (**Nentwich & König, 2014**). In academia, the most frequent scenario is probably the lowest degree of user participation; consenting to a very rudimentary profile with random contacts and infrequent online activities. The second-most frequent case allows creating a slightly more detailed profile, such as a more complex personal homepage (hence the name *digital calling card*) but no online interaction. Scholars involved in networking passively are characterized by using the network in irregular intervals for other (previously known) members, reacting to automated suggestions to contact other users, and sporadic communication with other members. A much higher level of involvement leads to active networking and communication, such as being regularly online, using further services, participating in group forums, and actively searching for potential networking partners. The highest level of the online presence and activities, labelled as*cyberentrepreneurship*, is when the individual does not only actively participate in the online activities, but does it for the benefit of others too by moderating forums, administering groups, taking control of institutional profiles, and giving feedback to the site developers. This is the rarest form of participation of researchers in the scholarly social media (**Nentwich & König, 2014**).

4.1. The librarian's role: scholarly selfies classified

Scholarly social media applications can be best understood and differentiated via a typology based on their main focus and primary goal. A classification offered previously points out the differences in five major areas: profiles related to*search-engines* (Google Scholar Citations and Microsoft Academic Search),*filesharing* (academia.edu, BEPress Selected Works, figshare, and ResearchGate),*researcher IDs* (ORCID, ResearcherID), *citation management* (Mendeley, CiteULike), or *specific research areas* (MyNCBI, Social Science Research Network) (**Dudás, Ward, & Bejarano, 2014**; **Hajnal Ward et al., 2014**).

Added here are two more evaluative dimensions: social media interoperability and author control. These dimensions better describe the variety of functionalities, along with the choices currently available for researchers in terms of enhancing their online presence. It should be noted that the features with which the individual applications are tagged are not mutually exclusive, and one application belongs to several categories. For this classification, the primary function has been used, to provide guidance for those interested in the main characteristics of scholarly social media opportunities as well as for those wishing to make their choices for their online presence. A special distinction across the platforms should be made in advance: while in most cases the researcher is responsible to set up a profile, many of these platforms feature built-in, or better to say, pre-fabricated profiles, without notifying the author. These dormant profiles might have been created on the basis of papers uploaded by co-authors or collected and indexed by the database.

4.2. Traditional profiles meeting social media

Researcher profiles have existed in subscription databases and on institutional websites for a long time. A simple faculty homepage, updated once or twice a year, can be a good starting point to create one's online presence and place the researcher's scholarship into context. Without dynamic features and the most current information, these pages fulfil the bare minimum of a scholarly profile. Some universities have started to use their home-grown applications to provide their faculty with a more dynamic version, which can be updated by the faculty member more easily. Instead of having to wait for a webmaster to upload content, the researchers and instructors have full control over their profile, such as in the Rutgers Faculty Survey by Rutgers, The State University of New Jersey. This is a questionnaire-type database, which has the capabilities of generating output in all required forms for reappointment, promotion, as well as a CV and a website for the faculty member. The final text is based on the data entered by the faculty member, and it is in their best interest to keep it current to meet all requirements in the next promotion cycle.

Another similar advancement is the result of the new biosketch format mandated by the National Institutes of Health (NIH). The biosketch, required for any grant-funded research proposal and subsequent research report, has always had restrictions in terms of length and content. Currently, NIH suggests that the fields in the form should be populated from the researcher's data generated from the MyNCBI profile to ensure that it is compliant with all policies. The form also features a link to the researcher's MyNCBI bibliography. Maintained by the National Center for Biotechnology Information of the U. S. National Library of Medicine in Bethesda, MD, the MyNCBI bibliography not only pulls open access articles published by the author from PubMedCentral, but it also communicates with the electronic submission systems via NIH and eRA Commons. The latest addition to the group is SciENcv (https://www.ncbi.nlm.nih.gov/sciencv/), a researcher biosketch profile service, which acts as the intermediary between the MyNCBI bibliography and the biosketch for the grant proposal required by the NIH. With the capability of adding bibliographic entries manually, now the authors can keep their bibliography current, including records not automatically pulled from NIH-compliant databases, as well as formats such as book chapters or less traditional scholarly output.

In Europe, there is a growing need for comprehensive regional bibliographies that cover full outputs of scholars and organizations, including all research fields, all publication types, and all languages (**Dudás et al., 2014**; **Engels, Ossenblok, & Spruyt, 2012**; **Sivertsen & Larsen, 2012**). In the social sciences and humanities (SSH), the two traditional bibliometric subscription databases (Web of Science and Scopus) focus on scholarly publications in English. However, "research outputs include not just those articles published in international journals, but also articles in national journals, academic book chapters and books, books aimed at a more popular audience, monographs, reports in the 'grey literature', and non-published outputs from fields such as the performing arts" (**Martin et al., 2010**, p. i). In particular, these gaps led to a national academic bibliometric database in Hungary, called the Database of Hungarian Scientific Works (Magyar Tudományos Művek Tára, MTMT) founded in 2009 with the aim of providing a current, comprehensive, and authentic overview of the national publication output and its impact. MTMT is a national research publication documentation system, a specific mixture of social media and shared, traditional electronic bibliographies. Basic bibliometric indicators, such as the total publication count and citation count, statistics by domestic/foreign languages, items by home country/ abroad, number of articles in scientific peer reviewed journals/non-scientific journals, and other features are calculated, taking into consideration field differences in publication and citation norms. Interconnected with other evaluation systems, MTMT is the prime source for basic bibliometric data in Hungary. MTMT author profiles can be linked to other web-based profiles, such as the author's institutional home page, Web of Science via Researcher ID, Scopus, ORCID, Google Scholar, or ResearchGate. They can also be connected to other local or regional systems featuring similar academic profiles, such as the portal of Hungarian Academy of Sciences and Hungarian Doctoral Council.

The librarian's role in the above instances is to keep abreast of the new advancements, whether related to one's home institution, a funding agency, or any administration, and experiment with workflows on the new platforms with the help of a handful of researchers, who are willing to experiment and share profile data. Once comfortable with the platform, the new role the librarian can assume might include providing guidance and assistance, and even training for research assistance to benefit from these applications.

4.3. Engaging in the (scholarly) web: author control

Scholarly social media profiles can be placed anywhere on a dynamic scale based on the control the user has over them: from fully author-controlled to fully platform-generated. Author control is independent of other features. For example, search engine-based profiles and sharing applications (see later) might already have a dormant profile, which needs only to be activated by the user. On the other hand, some or all of the content can or cannot be altered, again, depending on the platform.

The level of control might be none, such as in most subscription-based - profiles (e.g., Scopus, ISI Highly Cited). In such a case, the author's page is generated by the application. There is only a little latitude to request modifications, such as merging several profiles accidentally created for the same author (for example, due to name discrepancies). Most scholarly social media applications tend to require the researcher's initiation to set up a profile with the appropriate credentials, which is almost immediately populated by the publications attributed to the author within the database, such as Google Scholar Citations, ResearchGate, Academia.edu, etc. The bibliographic entries are added either from the result of crawling the scholarly web, such as in Google Scholar Citations and Microsoft Academic Search, or pulled from existing records either collected and indexed by the database or uploaded by co-authors with existing profiles, such as in ResearchGate, MyNCBI, Academia.edu, etc.

The highest level of author control is provided with the potential of adding virtually any format to the profile, such as in ResearchGate, which greatly promotes sharing alternative scholarly output. In addition to the traditional peer-reviewed papers, for example, PowerPoint presentations can also be uploaded and assigned a digital object identifier (DOI) in the system. Scholarship is interpreted in the broadest sense, yielding to alternative ways of evaluation as suggested by altmetrics, and the author has full control over the content to be uploaded, including the decision to provide full text.

The ease of full-text uploads provides a significant contribution from the librarians to the profile-creating process, namely, raising awareness about copyright issues, including restrictions, journal policies, fair use, and professional courtesy, since most platforms do not and cannot verify copyright ownership. The librarian can also help with the settings of notifications to prevent annoying messages inundating the researcher's mailbox, since by default, most of the platforms feature an opt-out notification system.

4.4. Searching the (scholarly) web

Researcher profiles have the potential to provide an overview of the scholar's work by listing publications and highlighting major achievements. Search engines crawl the entire web, including research-related pages, thus making scholarly profiles based on search engines the easiest way to find out about a scientist's work. With its extensive database of scholarly publications, Google Scholar (GS) showed a great potential as a citation index at its inception (**Noruzi, 2005**), and has been used successfully to locate known items and to review the literature on a topic. The same database serves as foundation for scholarly profiles, available after registration, for virtually anyone who has ever published. Launched in 2012, GS offers one of the simplest options to create a profile. After setting up a plain profile page with affiliation, contact information, and picture, as well as a link to the researcher's home page, GS populates the profile with the author's articles, and the related metrics, such as citations to the articles, h-index, and some other metrics and graphs. The bibliographic records are linked to the source, which may lead to the full-text via the user's own article-linker software, if subscribed to the particular journal. This is all done for the researcher and requires just a small amount of time to set up, though GS is not free from problems (**Jacsó, 2012**).

The same applies to the second platform based on a search-engine, Microsoft Academic Search, which offers a pretty profile page, pre-fabricated with basic metrics and charts. However, what the author gains at the setup is easily lost while trying to consolidate data, correct items wrongfully attributed to him or her, and add missing items. This application claims to be a research project created to follow information seeking in academia, and has not been updated (**Orduna-Malea, Ayllon, Martin-Martin, & Lopez-Cozar, 2014**). GS tends to overinflate researchers' scholarship, while Microsoft Academic Search is just the opposite. A common feature, in our experience, is the low count of non-English publications. Hungarian LIS papers are rarely found. Although there is an option to add articles manually, it is very time consuming and beats the purpose of a search-engine based application in the first place (**Jacsó, 2011**). GS and Microsoft Academic Search both could be used for evaluation proposes only if applied along with other citation indexes as a way to supplement that information (**Ortega & Aguillo, 2014**).

These two offer the least amount of control over the researcher profile, and as such, result in the least accurate data, causing the most frustration to the researcher with the over- or underinflated metrics. Apart from requesting the consolidation of the multiple profiles, there is no chance to correct a Microsoft Academic Search profile, which seems to be futile in the absence of updates anyway. The role of the librarian is to assist the authors to set up the basic profile in Google Scholar Citations and show how to manually add missing publications. It should be pointed out that the more accurate the data input, no matter how laborious and strenuous the process is, the closer to reality the metrics will be. Since researchers tend to prefer the inclusion of book chapters on their publication list, done so by GS, the librarian can also point them to manageable applications, where the data can be corrected and manipulated to compute the correct metrics, such as Publish or Perish and Scholarometer.

4.5. Sharing research

With a mission to expedite research all over the world, file-sharing systems, such as ResearchGate, academia.edu, or figshare, have shown the fastest growth and the greatest potential to create virtual research communities. Created by scientists for scientists, these applications are also promoting Open Science, since they wish to provide fast, free, and open access to research results by taking advantage of networked information environments and collaboration. The modern technologies behind these applications ensure that all file formats can be uploaded quickly and as simply as possible, then stored in a secure and easily accessible, cloud-based repository, while they also offer a variety of collaborative tools and even APIs to make the best use of the content. ResearchGate and academia.edu were each used by one fifth of the respondents according to a survey about using scholarly social media (**Haustein et al., 2014**).

The two main platforms, academia.edu and ResearchGate (researchgate.net) were launched in 2008, while figshare was first launched in 2011. Academia.edu was founded by an Oxford University philosopher as an academic social network site and currently has twelve million registered users. Part of its original purpose was to connect authors to readers so that it would be easy to send a query on a paper that had just been read (**Mangan, 2012**). Located in Cambridge, MA, ResearchGate has nearly four million members, after reaching three million in June 2013. ResearchGate targeted mostly the individual researchers at the beginning, and now expands to specific communities of users (**Gewin, 2010**), which explains the communication, collaboration, and sharing functions. Business investments show that the world outside academia also keeps an eye on scholarly social media: ResearchGate received over 35 million dollars from investors including Bill Gates (**Van Norden, 2014**). It should be noted that, similarly to other social media applications, these numbers show the registered users with profiles in various phases of completeness, rather than ones who actively use these sites. Figshare, advertised as "a community-based open data platform for scientific research," was first launched in January 2011 and relaunched after receiving support from Digital Science, a new division of Macmillan Publishers. Figshare offers unlimited storage space for data that is made publicly available on the site with full control over the management of their research (**Open data platform figshare..., 2012**).

In addition to monitor the trends and preferences within a scholarly community and show potential benefits, the librarian's role can be multifaceted. As is the case with any database, social media platforms are only able to generate metrics based on their own data. ResearchGate, for example, keeps track of download counts and citations; it even computes its unique number to indicate the author's scholarly impact, the so-called RG score, but all figures are calculated only from its own database. In other words, the author's citation count will rise if the citing article is in the RG database. This is not any different in the case of the subscription databases either, however, a social media database seems to be more unpredictable and contingent on its contributors, who may or may not favour a particular database. To promote inclusion, these platforms also make suggestions to add publications, invite authors and collaborators, and similarly to other, non-scholarly social media applications, they tend to push information in as many ways as possible. The librarian can bring these facts and features to the researcher's attention and help optimize settings and preferences.

4.6. Managing research: citation management as added service

Proprietary citation management applications, either web-based or desktop, such as EndNote, RefWorks, BibTex, and Reference Manager, now compete with free choices such as Zotero or CiteULike. A special configuration of scholarly social media is the application that combines all of the advantages mentioned so far with the functionalities of social bookmarking and citation management on the same one platform.

Established in 2007, the originally London-based and free Mendeley now links several, if not all, functionalities offered by other scholarly social media platforms within the same application, including an author profile, citations and bibliographies, social networking features (newsfeeds, comments), full-text search across papers, usage-based readership statistics, and public groups to share reading lists (**Zaugg, West, Tateishi, & Randall, 2011**). As a result, it has become possible to use a single platform for searching the scholarly literature, organizing and sharing search results, reading and annotating the articles, writing and sharing drafts with co-authors for comments, and creating the final text with references. Mendeley offers an option to set up private groups to collaboratively tag and annotate research papers, and there are apps for various devices, such as the Mendeley Desktop app, iPhone app, and iPad app. It is also possible to migrate content to and from services, such as Mendeley to ORCID. The platform boosts over three million users now enjoying either its free, basic version or the premium fee-based platform. Mendeley is considered important in indicating interest in the research from the readers' angle (**Li & Thelwall, 2012**). The fact that Elsevier purchased Mendeley in 2013 hints that there is a market for complex software applications in research that also include social media-inspired functions. Readership data from Mendeley is useful to capture knowledge transfer across scientific disciplines, since it shows impact earlier than citation counts. (**Mohammadi & Thelwall, 2014**). With its nearly eight million articles, CiteUlike (citeulike.org) is competitive with Mendeley, although the correlation between the bookmarking data and citation counts have been found more appropriate for research assessment with data from Mendeley (**Li & Thelwall, 2012**).

Librarians have long been conducting instructional sessions on how to use reference management systems. Providing training in Mendeley or demonstrating the advantages of CiteULike could be added to their responsibilities in an academic library setting. Librarians should also be part of university-wide discussions about whether to adopt one free system over another, or decide on the campus-wide premium subscription to Mendeley from Elsevier.

4.7. Disambiguating research: ID

The problem of a unique and general identifier for researchers is not new (**Foley & Kochalko, 2012**; **Haak et al., 2012**). Proprietary databases created their own identifiers to disambiguate authors, such as ResearcherID by Thomson Reuters or the Scopus Author Identifier. Each scholarly social media platform has its own identifier, such as Google

Scholar ID, ResearchGate ID, etc. A promising 2012 initiative, called ORCID (Open Researcher and Contributor ID), available at orcid.org, offers a solution to automate the collection of one's research output and related metrics via a 16-digit code, independent of providers, platforms, or applications.

ORCID was founded in 2012 as a non-profit organization comprised of publishers, funders, and institutions like Nature Publishing Group, Wellcome Trust, and Cornell University. Similarly to a DOI, the digital personal identifier is unique, and the ORCID account can be connected with other applications, platforms, and websites including MyNCBI, the Hungarian MTMT, Web of Science, figshare, and Impactstory, among many others. Once connected, information can be pushed back and forth between them, thus allowing to automatically import the same information to multiple places, rather than having to enter the same information over and over. Over 1 million researchers have ORCID IDs so far (**Foley & Kochalko, 2012**; **Haak et al., 2012**). Name variations during a researcher's career or due to diverse methods of data entry across platforms lead to inaccurate or multiple author profiles within the same database or social media platform, and as a result, miscalculated metrics and inaccurate representation of one's scholarly impact (**Hajnal Ward et al., 2014**). ORCID has a great potential to consolidate name variations across scholarly social media and traditional author profiles.

Although ORCID is still a relatively basic service with gaps in its coverage, librarians should raise awareness of the potential. Even though one cannot edit incorrect entries or export profile information everywhere, it has been integrated with a major researcher profile system, the MyNCBI, which serves as the basis of the new NIH-mandated biosketch with its SciENcv application.

5. MAKING SENSE OF SCHOLARLY SOCIAL MEDIA: THE ROLE OF THE LIBRARIAN

Emerging scholarly social media applications, alternative metrics, scholarly selfies, open science, open access – these are just a few new features academic librarians have come to embrace recently. Already involved in both bibliometrics and social media, librarians seem to have a unique opportunity to monitor new trends in the evaluation of scholarly output across the disciplines and follow major social media platforms with scholarly performance indicators. Many librarians got involved in the scholarly social media early either as researchers themselves or as reference on how to set up the accounts and create their profiles. Libraries, along with higher educational and research institutes, play a key role to open science as they pass their own Open Access initiatives, such as the Scholarly Open Access at Rutgers (**Rutgers Open Access Policy, 2012**.)

Researcher profiles on social media platforms represent a new and exciting way of promoting one's scholarship. In concert with the acceptance of unorthodox research output and the appreciation of new metrics to represent impact, the individual's "scholarly selfies" have gained importance through the options the hosting platforms offer to improve the research and publication process. In addition to listing one's traditional publications, such as journal articles and book chapters, the researcher profiles suggest considering non-traditional results and formats, such as data sets, codes, negative results, grey literature, blogs, and annotations. The platforms provide great opportunities to engage in research in a new way: communicate and collaborate with colleagues, discuss scientific topics, find co-authors, and so on. The exchange of full-text publications is often part of these sites, whether uploaded or requested as professional courtesy. The researcher profiles, with publications, data, and scholarly discussion, are open and visible to the world, which means that not only the privileged have access to science. The researcher and the host institution will benefit from a greater visibility and improved discoverability. The profiles provide data for altmetrics, and as such, serve complementary metrics to traditional citation analysis, representing the scholar's work in a more complete way. Scholarly selfies in mass also have an impact on the greater scientific community in terms of making science more transparent and open, disclosing research results faster and for broader audiences, and in the long run, speeding up the advances in science.

The authors investigated each scholarly social media platform on its own merit, and assessed their potential use for their particular audiences. Organizing and assessing the most popular platforms resulted in a better understanding of the advantages of one application over the other when creating academic profiles, or "scholarly selfies" for a particular purpose. Reviewing the applications would not have been possible without the active participation of the researchers, since many of these platforms target a specific field of science or are related to grant funding. Working with both highly-cited and early-career authors, it became discernible that many applications had dormant or pre-fabricated profiles with an uncontainable digital persona. This means that even if researchers were reluctant to take control over their online presence, they could not prevent their digital persona potentially misrepresenting their academic achievements. Ignoring scholarly social media can be detrimental to scholars and might lead to unforeseen consequences, not only to them and their institutions, but also to their colleagues and students. Even though maintaining scholarly social media profiles is often considered as a distraction or even a complete waste of time, the platforms and tools have the potential to augment academic careers, if used appropriately (**Bik & Goldstein, 2013**).

This discovery served as a major revelation in terms of the role of librarians, and came with the responsibility to educate our faculty members. By furnishing scientists with sufficient information on the evolving new world of scholarly social media, they will be better positioned to make decisions about whether they wish to take the opportunity offered by social media. Academic librarians have a unique position to participate in these exciting new advancements in science by exploring the role of the librarian or information specialist in a technology-rich environment. The consensus is that researcher profiles are sorely needed, but it would be time-consuming to keep abreast of all new developments. It is not the best use of a scholar's talent to try to sort out scholarly social media or spend time on manual data input. Librarians, working with a variety of academics, have the chance to follow the trends of scholarly social media platforms and altmetrics, and play an active role in promoting this specific area of information literacy and raise awareness of copyright issues at the same time. The role of librarians ranges from pointing researchers toward the direction of some compelling evidence supporting why they need social media (**Bik & Goldstein, 2013**) to actually investigating the impact of their university's research in less traditionally established venues. With new services, such as navigating the scholarly social media, the librarian is probably the best person to convince those who refuse to use the library (**Cronje, 2013**).

Last but not least, librarians can also use all of the above mentioned platforms for their own scholarship to create a more accurate online persona. There are numerous tools available, with a special regard to librarians in academic libraries (**Izenstark, 2014**). The new skills and hands-on experience in techniques to build and optimize digital identities will result in being considered as an authoritative source in the ever-changing field of scholarly social media, in addition to enhancing visibility for both their libraries and themselves. Reaching the highest level of engagement in scholarly social media, librarians can become cyberenterpreneurs for the benefit of research and science.

NOTES

[1]"Scholarly selfie" is a term coined by Judit Ward for in-house use in 2014 based on the 2013 word of the year, but was popularized in conferences by the authors and was accepted in the academic librarian community.

[2]*Keeping Up with the Kardashians* is a popular American reality television series that has aired since 2007. The proper noun has gained a new meaning in American English. According to the Urban Dictionary, a "Kardashian" is "famous only thanks to physical attributes and free exposure via the internet".

REFERENCES

1. Abt, H.A. (2000). Do important papers produce high citation counts? Scientometrics, 48(1), 65–70.

2. ACUMEN Portfolio (2014). Guidelines for Good Evaluation Practice with the ACUMEN Portfolio. ACUMEN Consortium. Retrieved April 17, 2015, from http://research-acumen.eu/wp-content/uploads/ ACUMEN-Guidelines-Portfolio-TOC-document-v13.pdf

3. Barbour, K., & Marshall, D. (2012). The academic online: Constructing persona through the world wide web. First Monday, 17, n.p. Retrieved March 18, 2015 from http:// firstmonday. org/ojs/index.php/fm/article/view/3969

4. Bar-Ilan, J., Haustein, S., Peters, I., Priem, J., Shema, H., & Terliesner, J. (2012). Beyond citations: Scholars' visibility on the social web. Retrieved March 18, 2015, from http://arxiv.org/abs/1205.5611

5. Bik, H.M., & Goldstein, M.C. (2013). An introduction to social media for scientists. PLoS Biology, 11(4), e1001535. Retrieved March 18, 2015, from http://journals.plos.org/plosbiology/ article?id= 10. 1371/journal.pbio.1001535

6. Bornmann, L. (2014). Do altmetrics point to the broader impact of research? An overview of benefits and disadvantages of altmetrics. Journal of Informetrics, 8(4), 895–903.

7. Bornmann, L., Mutz, R., Neuhaus, C., & Daniel, H.D. (2008). Citation counts for research evaluation: Standards of good practice for analyzing bibliometric data and presenting and interpreting results.

Ethics in Science and Environmental Politics, 8(1), 93–102. Retrieved March 18,2015,

8. Boyack, K.W., Klavans, R., & Börner, K. (2005). Mapping the backbone of science. Scientometrics, 64(3), 351–374.

9. Crawford, M. (2011). Biologists using social-networking sites to boost collaboration. Bioscience, 61(9), 736–736. Retrieved March 18, 2015

10. Cronin, B. (2001). Bibliometrics and beyond: Some thoughts on web-based citation analysis. Journal of Information Science, 27(1), 1–7.

11. Cronje, J.C. (2013). Why I don't use the library. Proceedings of the Conference of the International Association of Scientific and Technological University Libraries. Purdue University. Retrieved March 18, 2015

12. De Bellis, N. (2009). Bibliometrics and citation analysis: From the science citation index to cybermetrics. Lanham, MD: Scarecrow Press

13. De Groote, S.L. (2008). Citation patterns of online and print journals in the digital age. Journal of the Medical Library Association: Journal of The Medical Library Association, 96(4), 362–369.

14. Dudás, A., Ward, J.H., & Bejarano, W. (2014). Tudományos "szelfi": Szerzői profilok az interneten [Scholary selfies: Your author's profile on the internet. In Hungarian.]. In 46th Annual Conference of the Hungarian Library Association, Sopron, Hungary. July 18, 2014. Sopron, Hungary

15. Egghe, L. (2006). Theory and practise of the g-index. Scientometrics, 69(1), 131–152.

16. Engels, T.C., Ossenblok, T.L., & Spruyt, E.H. (2012). Changing publication patterns in the social sciences and humanities, 2000–2009. Scientometrics, 93(2), 373–390.

17. Eysenbach, G. (2011). Can tweets predict citations? Metrics of social impact based on twitter and correlation with traditional metrics of scientific impact. Journal of Medical Internet Research, 13(4), e123.

18. Fenner, M. (2014). Altmetrics and other novel measures for scientific impact. In S. Bartling, & S. Friesike (Eds.), Opening science (pp. 179–189). Heidelberg, New York: Springer Open. Retrieved May 1, 2015

19. Foley, M.J., & Kochalko, D.L. (2012). Open researcher and contributor identification (ORCID). In Proceedings of the Charleston Library Conference. Purdue University. Retrieved March 18, 2015

20. Galloway, L.M., Pease, J.L., & Rauh, A.E. (2013). Introduction to altmetrics for science, technology, engineering, and mathematics (STEM) librarians. Science & Technology Libraries, 32(4), 335–345.

21. Gewin, V. (2010). Collaboration: Social networking seeks critical mass. Nature, 468(7326), 993–994.

22. Giglia, E. (2011). Academic social networks: It's time to change the way we do research. European Journal of Physical and Rehabilitation Medicine, 47(2), 345–349. Retrieved March 18, 2015

23. Goodier, S., & Czerniewicz, L. (2012). Academics' online presence guidelines: A four step guide to taking control of your visibility. Retrieved March 18, 2015 from http://openuct.uct.ac.za/sites/default/files/Online%20Visibility%20Guidelines.pdf

24. Gorraiz, J., Gumpenberger, C., & Schlögl, C. (2014). Usage versus citation behaviours in four subject areas. Scientometrics, 101(2), 1077–1095.

25. Gruzd, A. (2012). Non-academic and academic social networking sites for online scholarly communities. In D. Rasmussen Neal (Ed.), Social media for academics: A practical guide (pp. 21–37). Woodhead Publishing

26. Haak, L.L., Fenner, M., Paglione, L., Pentz, E., & Ratner, H. (2012). ORCID: A system to uniquely identify researchers. Learned

Publishing, 25(4), 259–264. =004e1238e1275c277b42573a6 7667 65534447b497675592f653b672c57582a72752d709b856003

27. Hajnal Ward, J., Bejarano, W., & Dudás, A. (2014). Tudományos szelfi: Szerzői profilok az interneten (Scholarly selfies: Author profiles on the inernet. In Hungarian). Library Review / Konyvtari Figyelő, 24(3), 290–304

28. Hajnal Ward, J., Stewart, M., Cox, J., Candon, P., & Cook, S. (2011). Gyakorlati bibliometria: A tudományos tevékenység értékelése könyvtári eszközökkel [Practical bibliometrics: Evaluating scholarly activities with library resources. In Hungarian]. Könyvtári Figyelő, 57(1), 107–132

29. Hall, N. (2014). The Kardashian index: A measure of discrepant social media profile for scientists. Genome Biology, 15(7), 424. Retrieved March 18, 2015, from http://www.genomebiology.com/content/pdf/s13059-014-0424-0.pdf

30. Hammarfelt, B. (2014). Using altmetrics for assessing research impact in the humanities. Scientometrics, 101(2), 1419–1430. doi: 10.1007/s11192-014-1261-3

31. Haustein, S., Peters, I., Bar-Ilan, J., Priem, J., Shema, H., & Terliesner, J. (2014). Coverage and adoption of altmetrics sources in the bibliometric community. Scientometrics, 101(2), 1145–1163.

32. Haustein, S., & Siebenlist, T. (2011). Applying social bookmarking data to evaluate journal usage. Journal of Informetrics, 5(3), 446–457.

33. Hicks, D. (2005). The four literatures of social science. In H.F. Moed, W. Glanzel, & U. Schmoch (Eds.), Handbook of quantitative science and technology research (pp. 473–496). Springer Netherlands

34. Hirsch, J.E. (2005). An index to quantify an individual's scientific research output. Proceedings of the National Academy of Sciences of the United States of America, 102(46), 16569–16572.

35. Hood, W.W., & Wilson, C.S. (2001). The literature of bibliometrics, scientometrics, and informetrics. Scientometrics, 52(2), 291–314.

36. Hoover, S. (2014). Reputation management: Evaluating journals for publication and impact. Developments in Business Simulation and Experiential Learning, 41. Retrieved March 18, 2015

37. Izenstark, A. (2014). Look good when you're googled: Creating and optimizing your digital identity. Library Hi Tech News, 31(9), 14–16.

38. Jacso, P. (2008). Testing the calculation of a realistic h-index in Google Scholar, Scopus, and Web of Science for FW Lancaster. Library Trends, 56(4), 784–815.

39. Jacsó, P. (2011). The pros and cons of microsoft Academic Search from a bibliometric perspective. Online Information Review, 35(6), 983–997.

40. Jacsó, P. (2012). Google Scholar author citation tracker: Is it too little, too late? Online Information Review, 36(1), 126–141.

41. Leydesdorff, L. (1998). Theories of citation? Scientometrics, 43(1), 5–25.

42. Li, X., & Thelwall, M. (2012). F1000, Mendeley and traditional bibliometric indicators. In E. Archambault, Y. Gringas, & V. Larivière (Eds), Proceedings of the 17th international conference on science and technology indicators (pp. 541–551). Montréal: Science-Metrix and OST. Retrieved March 18, 2015

43. Mangan, K. (2012). Social networks for academics: Proliferate, despite some scholars' doubts. Chronicle of Higher Education, 58(35), 1–7. Retrieved March 18, 2015, from http://chronicle.com/article/Social-Networks-for-Academics/131726/

44. Martin, B., Tang, P., Morgan, M., Glänzel, W., Hornbostel, S., Lauer, G., ..., Zic-Fuchs, M. (2010). Towards a bibliometric database for the social sciences and humanities – A European scoping project. Retrieved April 17, 2015 from https://globalhighered. files.wordpress.com/2010/07/esf_report_final_100309.pdf

45. Mas-Bleda, A., Thelwall, M., Kousha, K., & Aguillo, I.F. (2014). Do highly cited researchers successfully use the social web? Scientometrics, 101(1), 337–356.

46. Meho, L.I. (2007). The rise and rise of citation analysis. Physics World 20(1), 32–36. Retrieved March 18, 2015, from http://arxiv.org/abs/physics/0701012

47. Meho, L.I., & Yang, K. (2007). Impact of data sources on citation counts and rankings of LIS faculty: Web of science versus scopus and google scholar. Journal of the American Society for Information Science and Technology, 58(13), 2105–2125.

48. Mendez, J.P., Curry, J., Mwavita, M., Kennedy, K., Weinland, K., & Bainbridge, K. (2009). To friend or not to friend: Academic interaction on facebook. International Journal of Instructional Technology & Distance Learning, 6(9), 33–47

49. Moed, H.F. (2006). Citation analysis in research evaluation. Springer Science & Business Media

50. Mohammadi, E., & Thelwall, M. (2014). Mendeley readership altmetrics for the social sciences and humanities: Research evaluation and knowledge flows. Journal of the Association for Information Science and Technology, 65(8), 1627–1638. doi: 10.1002/asi.23071

51. Nentwich, M., & König, R. (2014). Academia goes Facebook? The potential of social network sites in the scholarly realm. In S. Bartling, & S. Friesike (Eds.), Opening science (pp. 107–124). Springer International Publishing

52. NIH biosketch sample document. Retrieved March 18, 2015, from http://grants.nih.gov/grants/funding/424/SF424R-R_biosketchsample_VerC.docx

53. Noruzi, A. (2005). Google Scholar: The new generation of citation indexes. Libri, 55(4), 170–180

54. Open data platform figshare relaunches with added functionality and unlimited storage. (2012). Retrieved March 18, 2015 from http://www.digital-science.com/press-releases/open-data-platform-figshare-relaunches-with-added-functionality-and-unlimited-storage/

55. Orduna-Malea, E., Ayllon, J.M., Martin-Martin, A., & Lopez-Cozar, E.D. (2014). Empirical evidences in citation-based search engines: Is Microsoft Academic Search dead? Retrieved March 18, 2015, from http://arxiv.org/abs/1404.7045

56. Ortega, J.L. (2015). Relationship between altmetric and bibliometric indicators across academic social sites: The case of CSIC's members. Journal of Informetrics, 9(1), 39–49.

57. Ortega, J.L., & Aguillo, I.F. (2014). Microsoft academic search and google scholar citations: Comparative analysis of author profiles. Journal of the Association for Information Science and Technology, 65(6), 1149–1156.

58. The Oxford Dictionaries word of the year is...(2013). Retrieved March 18, 2015, from http://blog.oxforddictionaries. com/ 2013/11/word-of-the-year-2013-winner/

59. Peters, M.A. (2014). Open science, philosophy and peer review. Educational Philosophy and Theory, 46(3), 215–219.

60. Piwowar, H. (2013). Altmetrics: Value all research products. Nature, 493(7431), 159–159.

61. Priem, J., Groth, P., & Taraborelli, D. (2012a). The altmetrics collection. PloS One, 7(11), e48753. Retrieved March 18, 2015, from http://journals.plos.org/plosone/article?id=10.1371/journal.pone.0 048753

62. Priem, J., Piwowar, H.A., & Hemminger, B.M. (2012b). Altmetrics in the wild: Using social media to explore scholarly impact. Retrieved March 18, 2015, from http://arxiv.org/html/1203.4745

63. Priem, J., Taraborelli, D., Groth, P., & Neylon, C. (2010). Altmetrics: A manifesto (v.1.0). Retrieved March 18, 2015 from http:// altmetrics.org/manifesto

64. Rinaldi, A. (2014). Spinning the web of open science: Social networks for scientists and data sharing, together with open access, promise to change the way research is conducted and

communicated. EMBO Reports, 15(4), 342–346. doi:10. 1002/embr.201438659. Retrieved March 18, 2015

65. Rodgers, E., & Barbrow, S. (2013). A look at altmetrics and its growing significance to research libraries. The University of Michigan University Library. Retrieved March 18, 2015 http://deepblue.lib.umich.edu/handle/2027.42/99709

66. Rutgers open access policy (2012). Retrieved March 18, 2015 from http://www.libraries.rutgers.edu/rul/scholarly_comm/Rutgers_OA_Policy_2012-11.pdf

67. Schreiber, M. (2008). An empirical investigation of the g-index for 26 physicists in comparison with the h-index, the A-index, and the R-index. Journal of the American Society for Information Science and Technology, 59(9), 1513–1522.

68. Sivertsen, G., & Larsen, B. (2012). Comprehensive bibliographic coverage of the social sciences and humanities in a citation index: An empirical analysis of the potential. Scientometrics, 91(2), 567–575.

69. Sud, P., & Thelwall, M. (2014). Evaluating altmetrics. Scientometrics, 98(2), 1131–1143.

70. Thelwall, M., & Kousha, K. (2014). Academia. edu: Social network or academic network? Journal of the Association for Information Science and Technology, 65(4), 721–731.

71. Thelwall, M., & Kousha, K. (2015). ResearchGate: Disseminating, communicating, and measuring scholarship? Journal of the Association for Information Science and Technology, 66(5), 876–889.

72. Thelwall, M., Haustein, S., Larivière, V., & Sugimoto, C.R. (2013). Do altmetrics work? Twitter and ten other social web services. PloS One, 8(5), e64841. Retrieved March 18, 2015

73. Torres-Salinas, D., Cabezas-Clavijo, Á., & Jiménez-Contreras, E. (2013). Altmetrics: New indicators for scientific communication in web 2.0.

74. Van Norden, R. (2014). Scientists and the social network. Nature, 512(7513), 126–129. Retrieved March 18, 2015

75. Vinkler, P. (2010). The evaluation of research by scientometric indicators. Oxford: Chandos Publishing

76. Weller, K. (2015). Social media and altmetrics: An overview of current alternative approaches to measuring scholarly impact. In I.M. Welpe, J. Wollersheim, S. Ringelhan, & M. Osterloh (Eds.), Incentives and performance (pp. 261–276). Springer International Publishing

77. Weller, K., & Strohmaier, M. (2014). Social media in academia: How the social web is changing academic practice and becoming a new source for research data. IT-Information Technology, 56(5), 203–206.

78. West, R., & McIlwaine, A. (2002). What do citation counts count for in the field of addiction? An empirical evaluation of citation counts and their link with peer ratings of quality. Addiction, 97(5), 501–504

79. What does altmetric do? Retrieved March 18, 2015, from https://www.altmetric.com/whatwedo.php

80. Woelfle, M., Olliaro, P., & Todd, M.H. (2011). Open science is a research accelerator. Nature Chemistry, 3(10), 745–748

81. Zaugg, H., West, R.E., Tateishi, I., & Randall, D.L. (2011). Mendeley: Creating communities of scholarly inquiry through research collaboration. TechTrends, 55(1), 32–36

Chapter 10

On the role of research data centres in the management of publication-related research data

Sven Vlaeminck[1], Gert G. Wagner[2]

[1]German National Library of Economics/Leibniz Information Centre for Economics (ZBW), Hamburg

[2]German Data Forum (RatSWD), German Institute for Economic Research (DIW Berlin), Max Planck Institute for Human Development, Berlin, and Berlin University of Technology (TUB)

ABSTRACT

This paper summarizes the findings of an analysis of scientific infrastructure service providers (mainly from Germany but also from other European countries). These service providers are evaluated with regard to their potential services for the management of publication-related research data in the field of social sciences, especially economics. For this purpose we conducted both desk research and an online survey of 46 research data centres (RDCs), library networks and public archives; almost 48% responded to our survey. We find that almost three-quarters of all respondents generally store externally generated research data – which also applies to publication-related data. Almost 75% of all respondents also store and host the code of computation or the syntax of statistical analyses. If self-compiled software components are used to generate research outputs, only 40% of all respondents accept these software components for storing and hosting. Eight out of ten institutions also take specific action to ensure long-term data preservation. With regard to the documentation of stored and hosted research data, almost 70% of respondents claim to use the metadata schema of the Data Documentation Initiative (DDI); Dublin Core is used by 30 percent (multiple answers were permitted). Almost two-thirds also use persistent identifiers to facilitate citation of these datasets. Three in four also support researchers in creating metadata for their data. Application programming interfaces (APIs) for uploading or searching datasets currently are not yet implemented by any of the respondents. Least common is the use of semantic technologies like RDF.

Concluding, the paper discusses the outcome of our survey in relation to Research Data Centres (RDCs) and the roles and responsibilities of publication-related data archives for journals in the fields of social sciences.

KEYWORDS

Research Data Management; Research Data Centres; Journals; Libraries Economics

1. BACKGROUND AND INTRODUCTION

In the social sciences (especially economics, political science and sociology) more and more researchers analyse data provided by official statistics or by specialised providers of research data (e.g., from the

ALLBUS at GESIS1 or from the SOEP at DIW Berlin2). In addition, relevant data may often also be purchased from companies like Thomson Reuters or Bloomberg.

Especially in economics, compared to other branches of empirical research, the compilation of own datasets is not common. A major exception is the field of experimental economics, where researchers often generate their own datasets in the course of investigations motivated by game theory.

Although a rising number of publications in almost all scientific disciplines are based on the analysis of datasets, there are few effective ways to effectively replicate or re-examine the results of an empirical article, to verify it, or to make it available for re-utilisation and to support scholarly debates.

Even research data, which — in principle — is publicly available, will not typically be archived (e.g., in a final working-file) with respect to the specific selection and adjustment procedures. Therefore, while replications are not necessarily prevented, they are extremely difficult in the cases of ambitious analysis based on specific data selections and calculations.

The current situation confronts both the scientific community and scientific infrastructure service providers, like libraries and research data centres, with multiple challenges. In addition to questions concerning data availability and incentives for sharing data, there exist also infrastructure challenges. In particular the roles and responsibilities of scientific infrastructure providers, e.g., research data centres (RDCs), for managing and operating a data archive that facilitates the replications of published research often are not clearly outlined.

The first part of our paper describes some of the problems that lead to poor replicability of social sciences research. Then our paper describes the outcome of desktop research and an online survey evaluating scientific infrastructure with respect to their potential services for the management of publication-related research data in the field of social sciences.

The conclusion of our paper discusses the roles and responsibilities of several stakeholders for operating data archives for scholarly journals. Experiences in other scientific areas are integrated in our suggestions for establishing data archives that are based on the complementary know-how of research data centres (RDCs) and libraries.

1.1. Why is social science research often not replicable?

According to the literature the following reasons for missing replicability may be mentioned:

- First, and most importantly, there is a lack of incentives for researchers to share their data with the community. The academic reward system does not honour the often time-consuming efforts of data sharing — in sharp contrast to publications, although *"[a]n applied economics article is only the advertising for the data and code that produced the published results"* (**Anderson, Greene, McCullough, & Vinod, 2008**, p. 101).

- Furthermore, social scientists may worry that data sharing could lead to personal disadvantages. Researchers who work up and share data with the community do not receive appropriate compensation, e.g., reputation, for their efforts and may even suffer in terms of academic career because data sharing takes time that cannot be spent on research. In addition, many researchers suspect others will "misuse" their data, for example with faulty interpretations or by using a dataset without due reference to the creator of the dataset (**Fecher, 2014**). Eventually, the legal status of research data with regard to data sharing is not sufficiently clear, which also leads to reservations in data sharing.3

- Few social science journals have currently implemented guidelines requiring their authors to provide their data and statistical computation codes (**McCullough, 2009**; **Vlaeminck, 2013**). So called "data availability policies" may, in some instances, oblige the authors of empirical research papers to supply the underlying data of their results and the code/syntax of their analysis along with the manuscript of the article. Those policies often are in line with the "replication standard" formulated by Gary **King (1995)**.

- Useful infrastructure components for the management of publication-related research data are rarely applied, which, in turn, prevents any uniform way of citing the underlying data. Available technical solutions like Dataverse,4 a powerful tool for managing and documenting publication-related research data, are adopted by only a few journals. In this context a critical point focusses on how professional research data centres handle research-related data and what kind of services, if any, they offer.

1.2. Do research data centres offer services for archiving publication-related research data?

Research data centres could actually be ideal institutions for managing publication-related research data published as attachments to articles within scholarly journals. These capacities originate from decades of expertise in the handling of social- and economic research data, from core-competencies in the creation and maintenance of metadata collected and tagged from surveys as well as extensive experiences in managing access to data (**Research Information Network, 2011**). **Cox and Pinfield (2013)** argue that librarians, in contrast, already often feel over-taxed with the multiple roles that they have in the various activities of their libraries. In addition, libraries may lack technical knowledge, domain-specific expertise and may also have limited personal experience in the common research processes. As such it may be difficult to position libraries as key players in this area. A loophole could be the close collaboration between libraries and research data centres to solve upcoming challenges, as **Christensen-Dalsgaard (2012)** suggests.

Therefore, the EDaWaX (European Data Watch Extended – www.edawax.de) project, funded by the German Research Foundation (DFG) 2011 to 2016, conducted a study evaluating if such services for publication-related research data are currently available from scientific infrastructure service providers like research data centres, libraries and archives. For this purpose a list of 46 scientific infrastructure organisations was prepared. It includes all German research data centres and data service centres accredited by the German Data Forum (RatSWD),5 research data centres organised within the Council of European Social Science Data Archives (CESSDA),6 the library networks in Germany as well as individual libraries and public archives.

Our investigation into the services provided by these data centres for managing publication-related research data in the social sciences is a hands-on approach to evaluate the possibility of cooperation between research libraries and research data centres. Therefore our study followed the suggestion of **Lyon (2012)** to develop *"a proactive approach to collaborating with disciplinary, national and international data centres ... for data deposit in such archives"* (p. 130).

In a first step, the websites of these organisations were examined with regard to potential services for storing and hosting publication-related research data. The ICPRS (Inter-university Consortium for Political and Social Research — University of Michigan) provides a publication-related archive7 that is used by numerous authors to deposit their publication-related data.8 NARCIS,9 a research information system located in the Netherlands, offers a specific service for publication-related data.10 DANS EASY,11 another service located in the Netherlands, can also be used to deposit such data in principle.12

However, desk research could not uncover other information needed for further analysis, which is why, in order to start a more detailed evaluation of potential services by these organisations, an online survey was conducted.

2. THE ONLINE-SURVEY

In October and November 2012 an online-questionnaire was sent to 46 organisations — among them 35 national and international research data and data service centres, 1 archive, 7 library networks and single libraries, as well as three other organisations (non-European research data centres). A satisfactory, especially when compared to average return rates of mail survey.

Due to the structure of the questionnaire, not all participating organisations responded to all questions, which explains deviations in the number of responses (Figure 1).

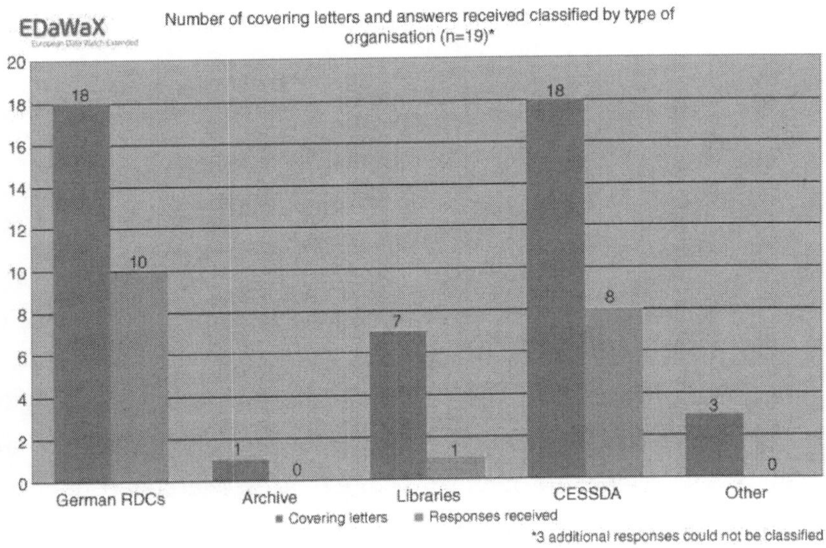

Figure 1. Covering letters and responses received.

Certainly more important than the return rate is the structure of respondents and non-respondents. The large majority of responses came from research data centres in Germany and Europe (86%). Significantly under-represented were respondents from German library networks and archives. The three non-European research data centres did not respond.

We can only presume that the library networks and the archive do not offer relevant services for research data management and, therefore, did not respond to our survey.

2.1. Empirical findings

Initially, the survey asked whether institutions would, in principle, host and store publication-related research data.13 In addition, the survey also asked whether organisations would also host and store (self-compiled) software components and the code of computation/syntax of statistical analyses. These three types of data are often part of empirical submissions to economic journals.14

2.2. Datasets

More than three-fourths of all organisations responding accept external datasets for storage (Figure 2). At the same time the lion's share of respondents reported that research data would only be accepted if certain criteria were met. Such criteria are subject to the specific competencies of many research data centres, but also to the specific regional/supra-regional or national competencies. Moreover, technical and organisational aspects (e.g., proper documentation, machine-readability, etc...) as well as legal concerns were cited as criteria. Approximately 74% of the respondents indicated that their organisations would host these types of data (Figure 3). If any criteria for hosting were mentioned, the subject-specific orientation of an institution was stated as main criterion

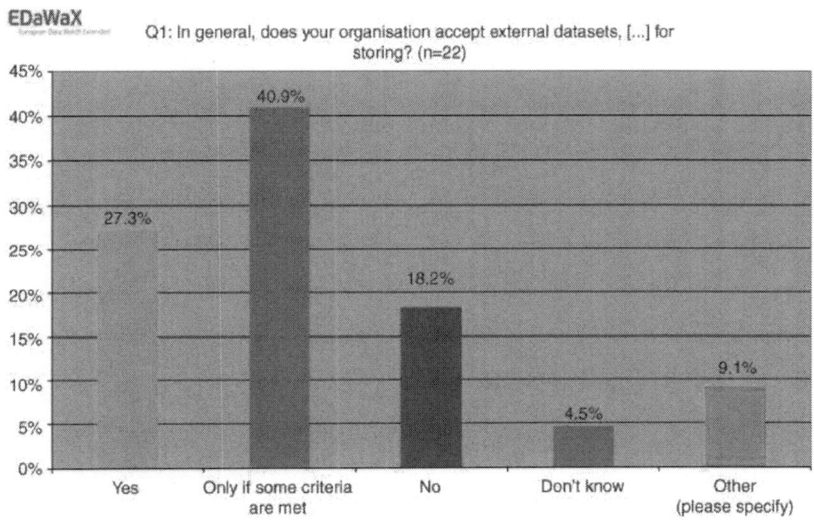

Figure 2. Acceptance of externally created datasets for storing.

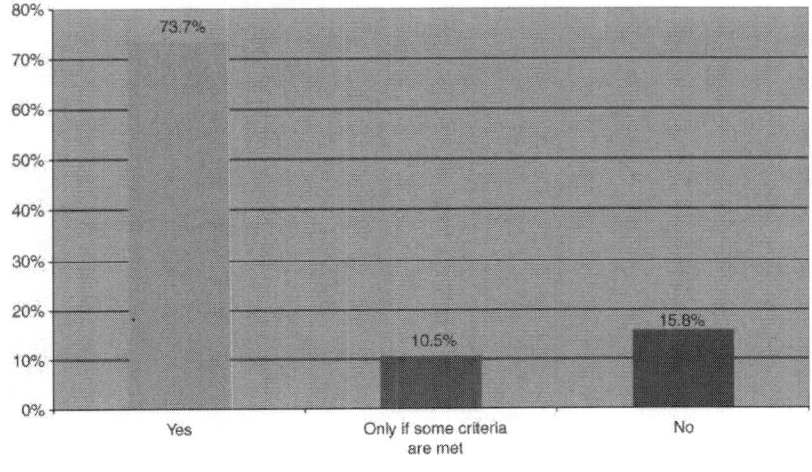

Figure 3. Hosting of externally created datasets.

2.3. Software

With regard to storing and hosting of (self-compiled) software components, which are often used for economic simulations, our survey indicates that just under a fourth of responding organisations accept storing and hosting software components without restrictions (Figure 4). Another 17% pointed out that they have criteria for assessing if software can be stored and hosted (e.g., *if essential for the analysis of the data*). Therefore, a gap exists in the availability of hosting and storing software components. Only a limited number of organisations offer this service.

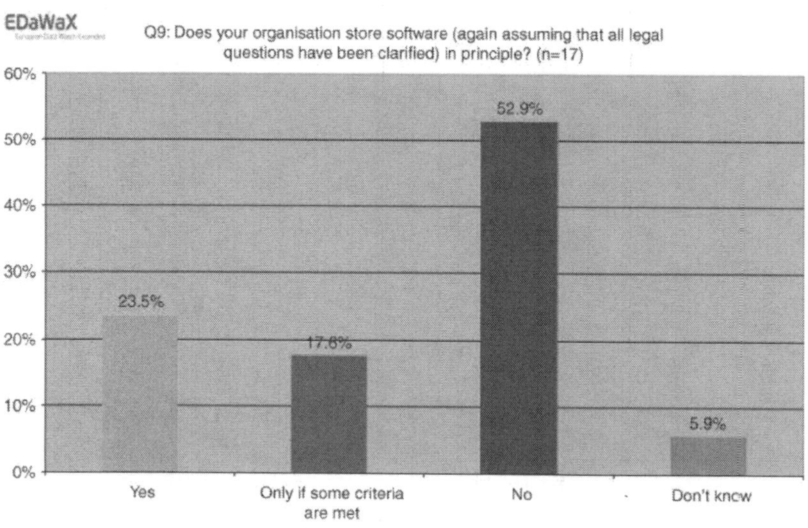

Figure 4. Storage of software components (e.g. used for simulations).

2.4. Code of computation/Syntax of statistical analyses

Almost 70% of the organisations responding offer options to store and host computation codes (Figure 5). However, a quarter does not do so at present and is not considering offering such services in the near future. One respondent also stated a criterion — noting that the storing and hosting of these data would only be useful in the case of derived variables.

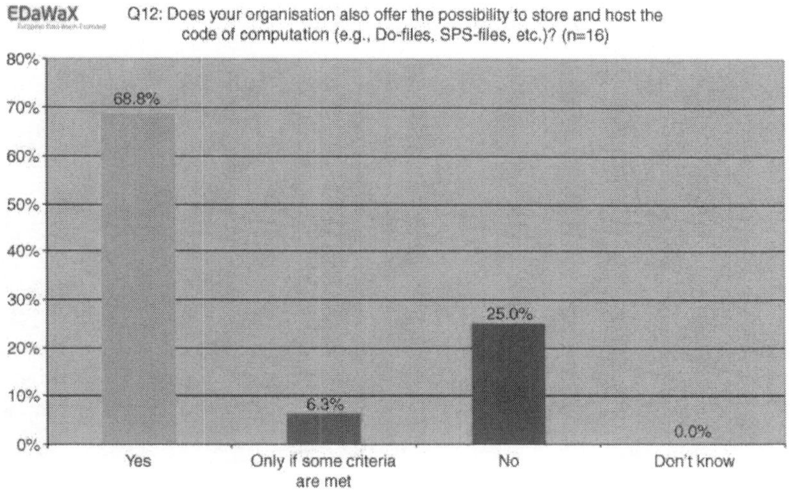

Figure 5. Storage and hosting of the syntax of statistical analyses.

2.5. APIs

Within our analyses we also examined the availability of application programming interfaces (APIs), which enable automated data exchanges. Our results show that less than half of all responding organisations have these interfaces at their disposal (Figure 6).

Most frequently APIs were mentioned as a device for data search (47%), followed by APIs used for uploading research data. Slightly more than a third (35%) of all respondents declared the availability of an API to analyse research data.

However, further analysis by EDaWaX shows that the reported interface consists only of searching and uploading interfaces on the respondents' websites. We were not able to find an API. Presumably, APIs in terms of external reading and writing accesses are by and large unknown among our respondents and not readily available.

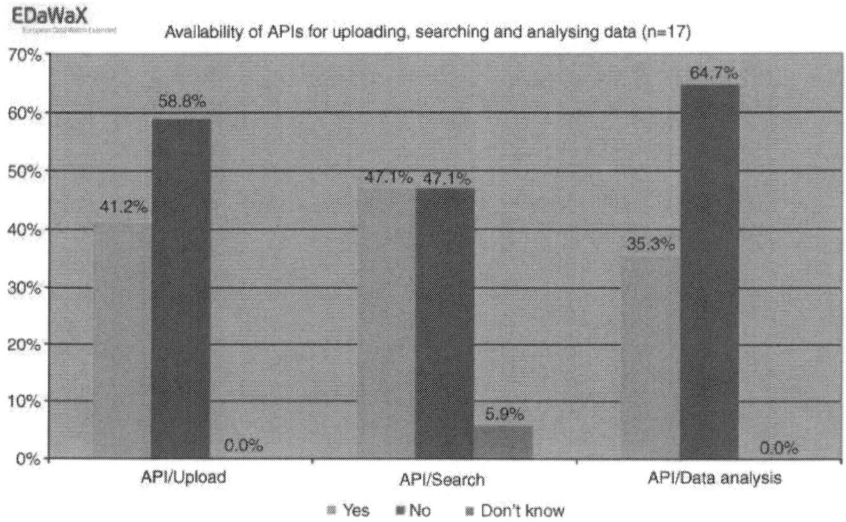

Figure 6. Availability of APIs.

2.6. Metadata schemata and the creation of metadata

2.6.1. Employed metadata schemata

We were also interested in the metadata schemata currently used by the organisations in their daily work. Our survey shows that more than 70% of the respondents use DDI (Figure 7). Other schemata like Dublin Core are rarely used (29%).15 All other metadata schemata are used rather sporadically.

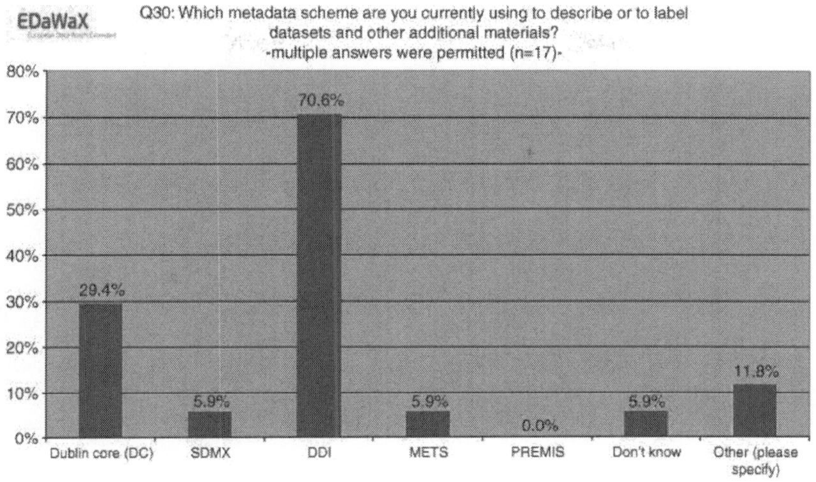

Figure 7. Metadata schemata currently in use.

2.6.2. Persistent Identifiers (PI)

In addition, we asked, whether organisations assign persistent identifiers (e.g., handle, DOI, URN, etc...) to datasets and other materials. The persistent identification of research data is an important issue, for instance because it enables researchers to cite datasets.

More than 56% of the organisations in our sample assign such identifiers by default, but almost a third do not (Figure 8). The persistent identification of research data is an important issue, for instance because it enables researchers to cite datasets.

2.6.3. Support of Semantic Web Technologies

In our survey we also examined the implementation of RDF (Resource Description Framework). RDF is a general method for conceptual description or modelling of information implemented in web resources. Among the organisations answering this question a minority of 6% claimed to use and disseminate RDF-files. Almost a quarter of all respondents did not specify whether their organisation uses RDF, which presumably indicates that RDF is largely unknown.

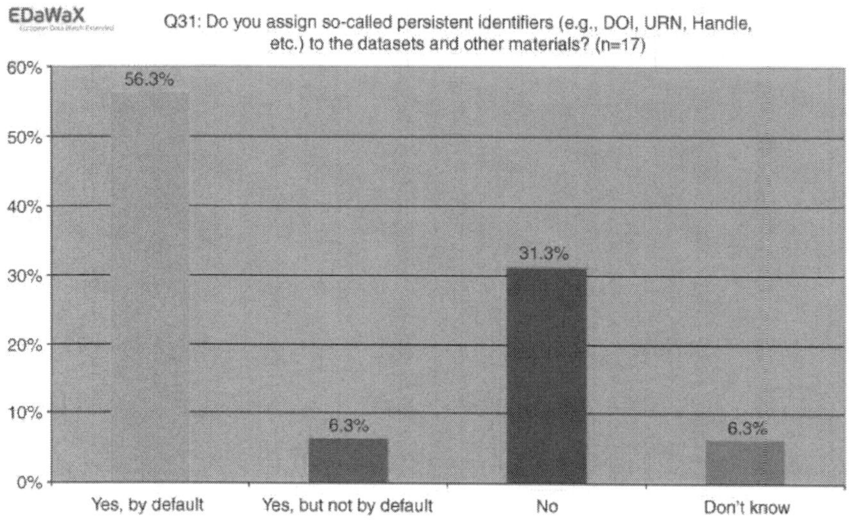

Figure 8. Assignment of persistent identifiers.

2.6.4. Support for creating metadata

Again and again, a critical issue regarding the reuse of research data is the quality of data documentation. Therefore, a matter of particular interest is whether respondents support researchers in generating metadata and, if so, how.

Our survey shows, that the majority (almost 65%) of all responding organisations do so (Figure 9).

Furthermore, we were keen to know whether this support is software-based — e.g., if there is a web frontend where researchers may type in the required information that is then converted into a standardised metadata schema.

We find that 36% of the respondents use this type of software-based support with researchers (Figure 10).

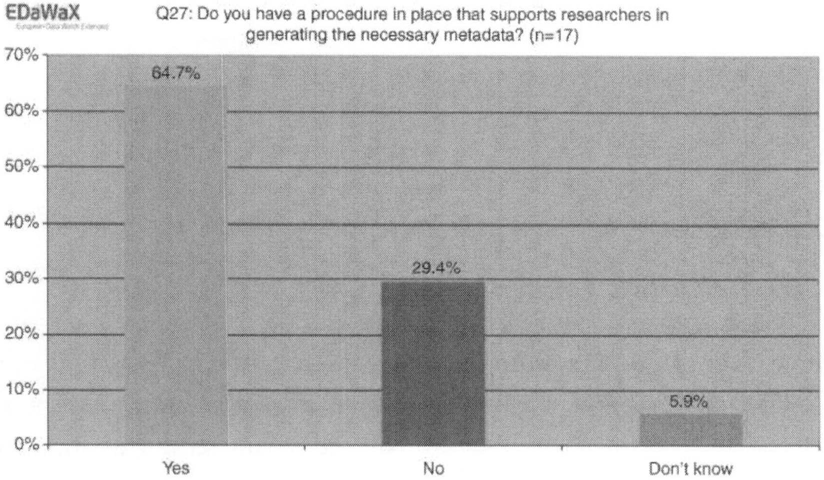

Figure 9. General support for metadata creation.

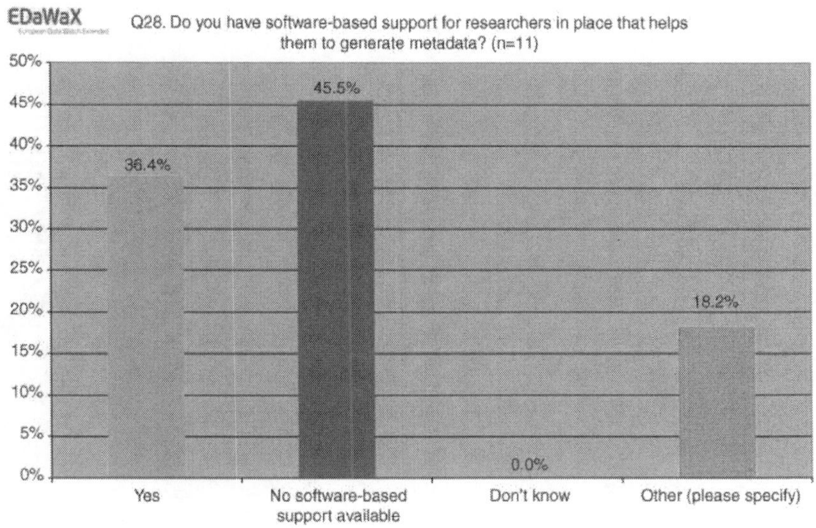

Figure 10. Software-based support for metadata creation.

There are a striking number of statements in the section *other*. Part of the *other*support for researchers, for instance, consists of written *data deposit forms*.

Our question regarding the software program names revealed that at least two institutions use Nesstar.16 Many organisations also use in-house solutions.

2.6.5. Digital long-term preservation

In our survey we wanted to identify to what extent the respondents' institutions have implemented specific measures for long-term research data preservation. Therefore we asked the respondents whether their organisations take specific actions for digital long-term preservation. Because format-migration is one of the dominant strategies for long-term preservation (**Harvey, 2012**), we suggested format migrations as one such method.

Our survey indicates that more than 80% of all organisations use these types of procedures (Figure 11).

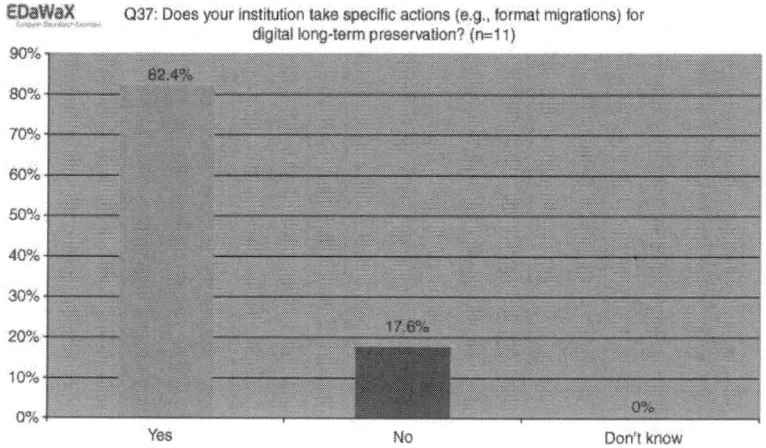

Figure 11. Long-term preservation of research data.

3. CONCLUSION AND DISCUSSION

Our study aims to evaluate if services for publication-related research data are available from data centres, libraries and archives. Based on existing services, our project defines roles and responsibilities for operating a publication-related data archive for journals in the fields of social sciences. This approach is in line with numerous recommendations put forth by European and national organisations as well as projects to interrelate research outputs with their underlying research data (**German Council of Science and Humanities, 2012**;**Kroes, 2012**; **Reilly, Schallier, Schrimpf, Smit, & Wilkinson, 2011**).

A question often arising in the context of linking data and publications is the discussion about stakeholders and their roles and responsibilities in the process (**Costas, Meijer, Zahedi, & Wouters, 2013**; **Lyon, 2007**). At first glance publishers appear to be the optimal stakeholders to perform the task of building up effective and efficient data archives because many publishers already host supplementary material for their journal articles. Therefore developing and implementing data archives, collecting and disseminating research data and metadata for datasets and other material could be an easy task for publishers. Therefore wouldn't it be a good idea to rely on the publishing industry? For answering the question we have to differentiate the role of academic publishers.

On the one hand, currently publishers do not see the need to implement data archives for journals on their own (**De Waard, 2012**). One reason might be that implementing and operating data archives raises the costs of publication. On the other hand, the availability of a data archive does not necessarily increase the number of journal subscriptions. Hence, the incentives to build up and operate data archives are not readily apparent to publishers.

In addition, questions of ownership and access conditions to archived research data could cause uncertainty for researchers, despite a publisher's announcement *"not to require any transfer of or ownership in such data or data sets as a condition of publication of the article in question"* (**STM & ALPSP, 2006**, p. 1).

Despite the fact that publishers do not operate data archives for their journals, they can nevertheless play an important role in the process of interrelating research data and publications. We already observe such collaborations in some scientific disciplines where publishers and data archives actively cooperate. In disciplines like the earth sciences, the e-infrastructure needed for storing and hosting research data in conjunction with appropriate documentations of the data has already been ongoing for quite some years. From a publisher's perspective linking research data and publications provides a benefit for their journals if the scientific outputs that are enriched with research data generate more citations (**Piwowar, Day, & Fridsma, 2007**). In addition, these links enable a more accurate research process and offer protection against scientific misconduct (**McCullough, 2009**).

Excellent examples of collaborations between publishers and data repositories include PANGAEA and Dryad. PANAGEA is the "data publisher" for earth and environmental sciences. It partners with Reed-Elsevier.17 Dryad is a non-profit repository for data underlying the international scientific and medical literature. It partners with numerous journals.18

Based on these experiences, the best solution is to implement and operate a discipline-specific data archive that gains importance by acquiring more and more data, which subsequently partners with publishers. The evolution and success of PANGAEA and Dryad underlines this approach impressively.

So, if it is not up to the publishers to run a data repository, other stakeholders come to the fore. In particular, research libraries and research data centres are the best positioned to take on the responsibilities of running such a disciplinary data repository. The results of our empirical investigation lead us to the conclusion that research data centres (RDCs) are likely the most relevant places to taking on the role of hosting and storing publication-related research data that is submitted to journals. RDCs already meet many prerequisites. In particular, the RDCs we analysed, in the broader field of social sciences, have much data handling experience. They are well trained in the storing, handling and documentation of these types of data as well as in taking appropriate measures for long-term data preservation.

Because RDCs currently do not comply with all requirements with respect to storing and hosting publication-related research data, collaborations between libraries and research data centres appear to be a promising way for establish such data archives: Libraries have the skills for managing publications. These include a dedicated knowledge of using authority files and multiple metadata schemata, in cataloguing information and providing this information to their discovery systems. Or as James L. Mullins, Dean of Purdue University's Library, describes it, *"Our ability to see structure to overlay on a mass of disparate 'parts,' as well as the ability to identify taxonomies to create a defined language for accessing and retrieving data is what is needed from us"* (**Baykoucheva, 2011**, p. 46). Unlike RDCs, it seems to be much more common for libraries to provide their stocks to their customers and to implement technical systems and the APIs necessary to do so.

According to **Pullinger and Wagner (2010)**, managing research data comprises of a mix of information that goes beyond the traditional separate realms of publications (the primary responsibility of national libraries), official records (the responsibility of national archives) and datasets (the responsibility of researchers themselves, statistical offices and RDCs) (**Pullinger & Wagner, 2010**, p. 3). In addition, **Cox and Pinfield (2013)** emphasize that scientific libraries often do not possess specialised units experienced in both IT-skills and knowledge in domain specific research data — a factor that hinders libraries' engagement in research data management. Establishing these departments takes time and costs money — often not attractive to scientific libraries during times of budget cuts.

Based on Lyon's suggestions to assign roles between libraries and RDCs (**Lyon, 2007**; adapted by **Vlaeminck, 2013**), we suggest the following tasks for the implementation of our project's pilot application of a data archive for economics journals, in which we strive to realise a workflow based on this division of complementary know-how.

In this distribution of tasks, ZBW — the Leibniz Information Centre for Economics — adopts the role of hosting and maintaining the metadata catalogue. Libraries then provide the technical implementation of APIs to other (library or research data) catalogues with the purpose of enriching and disseminating metadata. However one of the German RDCs, the research data centre of the Socio-Economic Panel (RDC SOEP), should take over the tasks of hosting, storing and preserving the data that has previously been submitted by editorial offices using the project's application.

By developing, implementing and operating a publication-related data archive for economics journals, both libraries and RDCs would help to ensure the validity of published economic research and to facilitate replications of these scientific outputs.

ACKNOWLEDGMENTS

The findings presented in this article have been achieved in the course of the EDaWaX (European Data Watch Extended, www.edawax.de) research project. EDaWaX is funded by the German Research Foundation (www.dfg.de). The institutions listed hereafter are involved in the project: The German Data Forum (RatSWD), the institute Inno-tec of the LMU Munich in cooperation with the Max Planck Institute for Intellectual Property and Competition Law (IMPRS-CI) as well as the German National Library of Economics / Leibniz Information-Centre for Economics (ZBW).

In addition to the authors the following persons are involved in the EDaWaX project: Professor Klaus Tochtermann (ZBW), Professor Joachim Wagner (Leuphana University, Lueneburg), Professor Dietmar Harhoff (IMPRS-CI and MCIER), Doctor Brigitte Preissl (ZBW), Patrick Andreoli-Versbach (IMPRS-CI), Doctor Frank Mueller-Langer (IMPRS-CI and MCIER), Olaf Siegert (ZBW), Ralf Toepfer (ZBW) and Doctor Hendrik Bunke (ZBW).

In particular the authors thank Adam Lederer (DIW Berlin) for valuable comments and suggestions for improving our English.

NOTES

[1]The ALLBUS (German General Social Survey) collects up-to-date data on attitudes, behaviour, and social structure in Germany. Since 1980 representative cross-sections of the population are surveyed by GESIS every two years using both constant and variable questions. Cf. http://www.gesis.org/en/allbus/allbus-home/.

[2]The German Socio-Economic Panel Study (SOEP) is a wide-ranging representative longitudinal study of private households in Germany, in which currently about 30,000 adults living in about 15,000 households are interviewed by TNS Infratest Sozialforschung on behalf of German Institute for Economic Research (DIW) Berlin each year. Cf. http://www.soep.de.

[3]Indeed, various reports and legal opinions on research data handling have been published in recent years, but it remains questionable if the uncertainty on the part of researchers has thereby been reduced [**De Cock Bruning, van Dither, Jeppersen de Boer, & Ringnalda, 2011**; **Guibault, & Wiebe, 2013**; **Häder, 2009**;**Hillegeist, 2012** (especially chapter A)].

[4]The website of dataverse can be found at: www.thedata.org.

[5]The website of the German Data Forum can be found at:http:// www.ratswd.de/eng/index.html.

[6]The website of CESSDA can be found at: http://www.cessda.org.

[7]In the meantime, ICPSR's publication related archive changed its name to "replication datasets."

[8]A list of all journals and articles in which data stored at the ICPSR-PRA is included are available athttp://www.icpsr.umich.edu/icpsrweb/ CPSR/ biblio/journals?collection=DATA.

[9]The Website of NARCIS, the National Academic Research and Collaborations Information System of the Netherlands can be found at:http://www.narcis.nl/about/Language/en.

[10]More Information on NARCIS can be obtained athttp:// www.narcis.nl/content/pdf/narcisflyer_en.pdf. According to DANS NARCIS currently contains more than 1,800 enhanced publications and 25,000 datasets.

[11]The Website of DANS EASY can be found on: https:// easy. dans.knaw.nl/ui/home.

[12]Useful information for instance was provided by **Polhout (2012)**.

[13]Respondents had the opportunity to inspect several examples of publication-related data submissions and their elements within the online questionnaire.

[14]The required elements depend on the type of research. The data availability policy of the American Economic Review (AER) – available athttp:// www.aeaweb.org/aer/data.php – exemplifies such requirements.

[15]Erratum: In our survey we considered XML as a metadata schema. XML is not a metadata schema but a markup language derived from

SGML. The purpose of XML is to define a set of rules for exchanging a wide variety of data. Therefore we no longer include these answers given by our respondents in our reporting.

[16]Website of nesstar, www.nesstar.com.

[17]Cf.http://www.reedelsevier.com/mediacentre/pressreleases/2010/Pages/elsevier-and-pangaea-take-next-step.aspx.

[18]Cf. http://datadryad.org/pages/integratedJournals.

REFERENCES

1. Anderson, R., Greene, W.H., McCullough, B.D., & Vinod, H.D. (2008). The role of data/code archives in the future of economic research. Journal of Economic Methodology, 15(1), 99–119
2. Baykoucheva, S. (2011). What do libraries have to do with e-Science? An interview with James L. Mullins, Dean of Purdue University Libraries. Chemical Information Bulletin, 63(1), 45–49
3. Christensen-Dalsgaard, B. (2012). Ten recommendations for libraries to get started with research data management. Final report of the LIBER working group on E-Science/Research Data Management. Retrieved January 30, 2014, from http://www.libereurope. eu/ sites/default/files/The%20research%20data%20group%202012%20 v7%20final.pdf
4. Costas, R., Meijer, I., Zahedi, Z., & Wouters, P. (2013). The value of research data – Metrics for datasets from a cultural and technical point of view. A Knowledge Exchange Report. Retrieved January 30, 2014 from www.knowledge-exchange.info/datametrics
5. Cox, A.M., & Pinfield, S. (2013). Research data management and libraries: Current activities and future priorities. Journal of Librarianship and Information Science. Published 'online before print', June 28, 2013

6. De Cock Bruning, M., van Dither, B., Jeppersen de Boer, C.G., & Ringnalda, A. (2011). The legal status of research data in the Knowledge Exchange partner countries. Retrieved December 10, 2013, from http://www.knowledge-exchange.info/Default. aspx? ID=461

7. De Waard, A. (2012). Linking data to publications: Towards the execution of papers. In: Uhlir, P. (Ed.), For attribution – Developing data attribution and citation practices and standards: Summary of an International Workshop (pp. 157–159). Retrieved January 30, 2014

8. Fecher, B. (2014). Data sharing angst – An insight to an ongoing research on data sharing in academia. Alexander von Humboldt Institute for Internet and Society. Retrieved January 30, 2014, from http://www.hiig.de/en/data-sharing-angst-an-insight-to-an-ongoing-research-on-data-sharing-in-academia/

9. German Council of Science and Humanities (Wissenschaftsrat) (2012). Empfehlungen zur Weiterentwicklung der wissenschaftlichen Informationsinfrastrukturen in Deutschland bis 2020 (No. Drs. 2359-12). Retrieved January 30, 2014, from http://www.wissenschaftsrat.de/download/archiv/2359-12.pdf

10. Guibault, L., & Wiebe, A. (2013). Safe to be open. Study on the protection of research data and recommendations for access and usage. Universitätsverlag Göttingen. Retrieved January 21, 2014, from http://webdoc.sub.gwdg.de/univerlag/2013/legalstudy.pdf

11. Häder, M. (2009). Der Datenschutz in den Sozialwissenschaften. Anmerkungen zur Praxis sozialwissenschaftlicher Erhebungen und Datenverarbeitung in Deutschland. RatSWD Working Paper Series, 90, Berlin. Retrieved December 10, 2013, from http://www. ratswd. de/download/RatSWD_WP_2009/RatSWD_WP_90.pdf

12. Harvey, D.R. (2012). Preserving digital materials (2nd ed.). Berlin: De Gruyter

13. Hillegeist, T. (2012). Rechtliche Probleme der elektronischen Langzeitarchivierung wissenschaftlicher Primärdaten, Göttinger

Schriften zur Internetforschung (8). Retrieved December 10, 2013, from http://webdoc.sub.gwdg.de/univerlag/2012/GSI8_ Hillegeist. pdf

14. King, G. (1995). Replication, replication. PS: Political Science and Politics, 28, 443–499. Retrieved December 10, 2013, from http://gking.harvard.edu/gking/files/replication.pdf

15. Kroes, N. (2012). Commission recommendation of 17.7.2012 on access to and preservation of scientific information [No. SWD(2012) 221 final]. Retrieved January 30, 2014, from http://dspace. utlib.ee/ dspace/handle/10062/34075

16. Lyon, L. (2007): Dealing with data – Roles, rights, responsibilities and relationships. Consultancy report. Retrieved January 30, 2014, from http://www.jisc.ac.uk/media/documents/programmes/digitalrepositories/dealing_with_data_report-final.pdf

17. Lyon, L. (2012). The informatics transform: Re-engineering libraries for the data decade. International Journal of Digital Curation, 7(1). doi:10.2218/ijdc.v7i1.220

18. McCullough, B.D. (2009). Open Access economics journals and the market for reproducible economic research. Economic Analysis and Policy, 39(1), 117–126

19. Piwowar, H.A., Day, R.S., & Fridsma, D.B. (2007). Sharing detailed research data is associated with increased citation rate. PLoS ONE 2(3): e308. doi:10.1371/journal.pone.0000308

20. Polhout, M. (2012). Deposit instructions for social and behavioural sciences. Retrieved December 10, 2013, from: http://www.dans.knaw.nl/sites/default/files/file/EASY/Deponeerinstructie%20MaGw%20UK%20DEF.pdf

21. Pullinger, J., & Wagner, G. (2010). On the respective roles of national libraries, national archives and research data centers in the preservation of and access to research data. Retrieved January 30, 2014, from http://www.econstor.eu/dspace/handle/10419/43600

22. Reilly, S., Schallier, W., Schrimpf, S., Smit, E., & Wilkinson, M. (2011). Report on integration of data and publications. Retrieved January 30, 2014, from http://www.alliancepermanentaccess.org/wp-content/ uploads/downloads/2011/11/ODE-ReportOnIntegration Of DataAndPublications-1_1.pdf

23. Research Information Network (2011). Data centres: their use, value and impact. A Research Information Network report. Retrieved December 10, 2013, from http://www.rin.ac.uk/system/files/ attachments/Data_Centres_Report.pdf

24. STM & ALPSP (2006). Databases, data sets, and data accessibility – views and practices of scholarly publishers. A statement by the International Association of Scientific, Technical and Medical Publishers (STM) and the Association of Learned and Professional Society Publishers (ALPSP). Retrieved January 30, 2014, from http://www.stm-assoc.org/2006_06_01_STM_ALPSP _Data_ Statement.pdf

25. Vlaeminck, S. (2013). Data management in scholarly journals and possible roles for libraries – Some insights from EDaWaX. LIBER Quarterly, 23(1), 48–79. URN:NBN:NL:UI:10-1-114595. Retrieved December 10, 2013, from http://liber.library.uu.nl/index. php/lq/ article/ view/URN%3ANBN%3ANL%3AUI%3A10-1-114595

Chapter 11

Developing data literacy competencies to enhance faculty collaborations

Don MacMillan

Librarian for Biological Sciences, Physics, Astronomy & Mathematics, University of Calgary, Taylor Family Digital Library, Calgary, Alberta, Canada

ABSTRACT

In order to align information literacy instruction with changing faculty and student needs, librarians must expand their skills and competencies beyond traditional information sources. In the sciences, this increasingly means integrating the data resources used by researchers into instruction for undergraduate students. Open access data repositories allow students to work with more primary data than ever before, but only if they know how and where to look. This paper will describe the development of two information literacy workshops designed to scaffold student learning in the biological sciences across two second-year courses, detailing the long-term collaboration between a librarian and an instructor that now serves over 500 students per semester. In each workshop, students are guided through the discovery and analysis of life sciences data from multiple sites, encouraged to integrate text and data sources, and supported in completing research assignments.

KEYWORDS

Data literacy; bioinformatics; collaboration; science librarians

1. INTRODUCTION

This case study describes how life science data was successfully incorporated into an information literacy (IL) program in order to better align instruction with student and faculty needs. The librarian and instructor had developed a successful workshop in Biology 311- Principles of Genetics, a second-year course in 2006 (**MacMillan, 2010**). The instructor noted that while most students in the genetics course took Biochemistry 393, Introduction to Biochemistry, in the following semester, they were not linking knowledge gained in one course to the other as well as she expected. Many undergraduate biology programs continue to teach genetics and biochemistry from different philosophical or pedagogical perspectives which inhibits students from fully understanding or appreciating the content that is common to both subjects (**Barrette-Ng & MacMillan, 2014**). These two courses, however, were taught by the same instructor, so there were opportunities to encourage students to develop a more integrated understanding of both disciplines. In 2010 the librarian and instructor collaborated on a second IL workshop for the biochemistry course that would foster transfer of knowledge between the two subjects, and introduce students to other domain-specific data resources. Students learned how similar data informed research in genetics and biochemistry, and were able to practice skills gained in navigating complex gene databases as they explored resources on proteins. The primary goal of lab was to remove some of the artificial barriers that exist between these two disciplines by having students examine the molecular basis of genetically inherited diseases. Specifically, students in Biology 311- Principles of Genetics, used relevant data sources to investigate the molecular basis of their genetically inherited disease topic, and in Biochemistry 393, during the winter semester, they examined the structural origins of the molecular basis of their disease topic with related data. Both the genetics and biochemistry information literacy sessions were developed with the same guiding principle: to introduce students to the benefits of using 'real-world' tools early in their academic careers. They were both structured to provide scaffolded practice exploring selected bioinformatics resources, followed by assignments that required integrating information from these resources with other course content to answer questions about the role of genes and proteins in genetically-inherited diseases. As with the first genetics session, working with the biochemistry class enabled the librarian to develop new data competencies and gain a deeper understanding of the two disciplines. This has contributed to a more robust and sustainable IL program that aligns more closely with faculty and student needs. The case study will provide background information on the resources included in the sessions, activities used to guide students in their exploration of these resources and suggestions for integrating bioinformatics data into information or data literacy instruction.

2. LITERATURE REVIEW

Researchers work with scholarly publications, data, and discovery tools in an integrated workflow. This suggests that academic librarians need to develop some familiarity and competencies with the scholarly communication patterns of the discipline in order to better integrate data and information literacy to educate the next generation of scientists. Providing an early introduction to bioinformatics data resources enables students to see science in action in fields where the published article may be almost secondary to the supporting data.Research data in these disciplines are in many cases more important than the articles they are linked or associated with (**Akers & Doty, 2013**; **Castelli, Manghi, & Thanos, 2013**). In a report by the National Academy of Sciences entitled *BIO 2010: Transforming Undergraduate Education for Future Research Biologists,* a key recommendation was the "need to involve students in working with real data and tools that reflect the nature of life sciences research in the 21st century" (as cited in **Ditty et al., 2010**, p. 1). Similarly, the *American Society for Biochemistry and Molecular Biology (ASBMB)* endorsed the use of bioinformatics and related data resources as a core competency for students in molecular biology courses (**Boyle, 2004**; **Voet et al., 2003**).

Bednarski, Elgin and Pakrasi (2005) were among the earliest to incorporate bioinformatics data into undergraduate courses. Bednarski developed an inquiry-based laboratory project which introduced undergraduate students to several key bioinformatics databases including the *Online Mendelian Inheritance in Man (OMIM),GenBank,* and *Protein Data Bank (PDB),* in order to study the effects of mutations on protein structure, function and human physiology. Students indicated they appreciated the investigative nature of these exercises and had a better understanding of how researchers develop hypotheses about genetic disease topics and share their scientific results. The authors also concluded that "the oral and written reports gave students a chance to practice their skills in communicating scientific information, and the collaborative nature and group work in the lab gave students opportunities to both defend their ideas and learn from their peers" (**Bednarski et al., 2005**, p. 216).

Further evidence of the benefits of incorporating bioinformatics resources within undergraduate information or data literacy instruction to enhance student understanding of molecular processes has been reported by a number of recent studies (**Badotti et al., 2013**; **Dymond et al., 2009**; **Weisman, 2010**; **Wightman & Hark, 2012**). Best practices for implementing similar programs have been delineated (**Badotti et al.,**

2013; **Bednarski et al., 2005**; **Miskowski, Howard, Abler, & Grunwald, 2007**; **Via et al., 2011**; **Weisman, 2010**), while others have discussed the programmatic role librarians can provide in supporting molecular biologists as they navigate the myriad data resources (**MacMullen & Denn, 2005**). Effective collaborations between librarians and the life sciences faculty have also been described in the literature (**Bowden & DiBenedetto, 2001**; **Brown & Krumholz, 2002**; **Callinan, 2005**; **Dinkelman, 2010**; **Ferrer-Vinent & Carello, 2008**; **MacMillan, 2010**; **Winterman, 2009**).

A common theme among many of these studies are the benefits associated with collaborating with the teaching faculty, including the potential to enhance librarians' domain knowledge and expertise in the subject area. Many of the authors also provide valuable guidance on how to use bioinformatics tools more effectively. Familiarity, if not expertise in using bioinformatics databases is now considered a critical competency in life sciences librarians, as articulated in two recent reports. One prepared for the *Association for Research Libraries* (ARL) examined the role of liaison librarians in research libraries and recommended more sustainable engagement strategies that shift the focus to what users do (e.g., research, teaching, and learning) rather than on what librarians do (e.g., collections and reference) (**Jaguszewski & Williams, 2013**). In the second study published by *Ithaka S+R*, the author posits that an "an emerging theme in the development of the liaison model is to shift the focus away from the work of librarians to that of scholars and to develop engagement strategies based on their needs and success indicators" (**Kenney, 2014**, p. 4).

In a study that examined the data information literacy needs of students and the research faculty, **Carlson, Johnston, Westra and Nichols (2013)** concluded that librarians can leverage teaching opportunities by integrating data into training and that a major "objective of this collaboration was to increase the capacity of librarians to engage with students and faculty" (p. 215). The authors of this study also determined that most researchers in this study thought that offering some type of data literacy was required for their students (**Carlson et al., 2013**). Other studies have also concluded that academic librarians are well positioned to support data literacy and data competencies (**Calzada Prado & Marzal, 2013**;**Gray, 2004**; **Schield, 2004**; **Stephenson & Schifter Caravello, 2007**).

3. CONTEXT

The development of the two IL sessions described in this paper rests on two contexts. The first is the rapidly changing information environment in the life sciences and the second is the institutional setting where the classes take place.

The completion of the Human Genome Project in 2001 resulted in a deluge of life science data, particularly from large-scale sequencing projects that generate data pertaining to the structural and functional aspects of genes and proteins (**Hey, Tansley, & Tolle, 2009**). The discovery and analysis of this type of data has been facilitated by the emergence of bioinformatics databases and the development of a robust data infrastructure including disciplinary repositories that provide quality control and sustainable preservation.

Furthermore, researchers in the life sciences are increasingly being mandated to adhere to funder and/or publisher mandates to disclose the results of their research, including data, in an appropriate disciplinary data repository so that findings from any publically funded research is accessible and reproducible to others (**Smith et al., 2011**; **Vickers, 2011**).

In addition to providing sustainable data preservation and discoverability, bioinformatics databases are a unique form of scholarly content, integral to the scholarly communication paradigm in the life sciences. They are now built into the work of researchers, journal publishers, libraries and funding agencies intent on making research results more widely available (**Bourne, 2005**).

The accessibility of life science data is further enhanced by the interoperability between key publishers and selected data repositories. For example, scientific publishers including Elsevier, Wiley, Springer and Nature are increasingly mandating that authors submit any data associated with a manuscript at an appropriate data repository such as he National Center for Biotechnology's (NCBI) *GenBank* and *Protein Data Bank* as a condition of publication. Increasingly, the article and associated data are linked bi-directionally which enabled students in this case study to begin their research on their disease topic from either the article or appropriate data repository.

While sophisticated, bioinformatics resources are relatively user-friendly, enabling inexperienced users to navigate and comprehend their potential if the right conditions are met (**MacMillan, 2010**). Fortunately, at the University of Calgary, the collaboration between the instructor and the librarian has met those conditions in developing workshops that not only introduce students to content-rich resources, but foster the integration of that content across disciplines.

The University of Calgary has a student population of approximately 31,000 students and offers over 200 undergraduate, graduate and professional degree programs. Information literacy support is provided for a variety of undergraduate programs including the biological sciences. When the author first worked with the biological sciences department, the main contact point with students was a first-year course with an assignment that introduced students to library resources in order to locate a peer-reviewed article on a research topic. However as students in first year did not need to do much research outside of the library assignment, this may not have been the most effective place for IL instruction, and the sessions were eventually dropped from the course in 2010. An instructor in the second year genetics course, however, did require her students to integrate information resources in their work, and so began the collaboration that is detailed in the case study. Information literacy instruction in these second-year courses is extended in third and fourth year in biochemistry and cellular biology courses where instruction includes reminders about bioinformatics content but focuses on bibliographic resources and information management.

The initial collaboration between the instructor and the librarian on an information literacy session for Biology 311-Principles of Genetics began in 2006, with the goal of providing a more authentic inquiry-based learning environment, and has been refined annually ever since to incorporate new resources, changes to interfaces and greater understanding of where students have the greatest difficulty. In 2010 this collaboration was extended to Biochemistry 393-Introduction to Biochemistry, so that students would better understand the content that is common to both disciplines by investigating the effects of mutations on different levels of protein structure, function, and genetically inherited diseases (**Barrette-Ng & MacMillan, 2014**). Lessons learned in developing Biology 311 were applied to the new class and both model authentic pathways that replicate researcher workflow, incorporate scaffolded steps and hands-on practice that leverage the interoperability of different bioinformatics resources. Both are assessed through short term assignments and larger projects that require further independent exploration of the resources.

4. CASE STUDY

This case study will outline each of the classes, and describe the information resources and activities. Details are provided so that others may adapt the work to their own contexts.

4.1 Biology 311-Principles of Genetics

This is a required course for biological sciences majors and covers topics such as Mendelian inheritance, sex determination, molecular genetics, and the structure and function of genetic material. The course is offered during the fall semester with enrollment in the fall ranging between 500–550 students. In mid-October students attend an information literacy workshop during their scheduled lab time. Typically three lab sections are combined for a total of 70–75 students who are accompanied by the Graduate Teaching Assistants for each lab. The session takes place in the library's computer classrooms which have 48 workstations and students generally work with their lab partners. The goal of this lab is the investigation of the molecular basis of genetically inherited diseases.

The teaching objectives for this lab include:

- Use PubMed to find a peer-reviewed article on a specific genetically-inherited disease topic.

- Locate a patent that pertains to a gene or mutation believed to cause a genetically-inherited disease.

- Use the *OMIM* database to determine the mode of inheritance of a genetically-inherited disease as well as the causative gene(s) and mutation(s).

- Use *NCBI's Gene* database to obtain data on gene architecture including mRNA and protein length.

Students work through a 20 question laboratory workbook during the session and then must revisit the resources to prepare and present a scientific poster using life science data that explains why some individuals inherit genetic diseases. During each IL session students are introduced to a series of exercises that focus on a specific resource used to answer a specific question and/or teaching objective listed above, related to the disease they have chosen to research. Students select their own disease topic which may enhance their motivation; many choose to research a disease that affects people they know which is likely to increase their engagement with the assignment.

The librarian provides a live demonstration of each resource using the *CFTR* gene that is implicated in Cystic Fibrosis. As each resource or bioinformatics database is introduced and demonstrated students are given the opportunity to practice with their own examples and additional time is available at the end of the session. The lab workbook provides a structured outline of the various bioinformatics tools including screen shots and explanatory notes that the students can use during and after the scheduled session. The questions in the workbook are progressively more complex and demanding, requiring students to knit together their understanding of the different resources as they complete the exercises. Following the session, each student is required to answer a series of questions each week about their chosen disease topic, which are integrated into a poster and oral presentation during the final week of term. The entire project is worth 6% of the overall course grade. A course related *LibGuide*(http ://libguides.ucalgary.ca/ content.php?pid=55723&sid=603862) with links and brief explanations of all of the resources used for this assignment provides additional structure for each session and a convenient entry point to each resource as students work on subsequent assignments and their posters. The librarian provides support while students work on the in-class assignment and is available for further consultation as needed before the assignments are due.

The resources in the BIOL 311 library session are described below in the order they are presented in the sessions.

Genetics Home Reference

(http://ghr.nlm.nih.gov/about)

This is the National Library of Medicine's web site for information about genetic conditions including the genes and chromosomes implicated with specific disease. It contains information on over 1000 genetically-inherited diseases or syndromes and over 1200 genes. Most students are able to retrieve data and background information on their disease topic such as the inheritance pattern of the disease, the name of the gene(s) implicated with their disease and the cytogenetic location or specific chromosome where a gene is located. As many diseases have more than one causative gene(s) or mutation, students were asked to investigate only one gene and mutation for their assignment and presentation.

Online Mendelian Inheritance in Man (OMIM)

(http://www.omim.org/)

OMIM is a comprehensive and authoritative database of human genes and genetic disorders hosted by the NCBI, and maintained by McKusick-Nathans Institute of Genetic Medicine at Johns Hopkins University School of Medicine and updated daily. *OMIM* utilizes NCBI's *Entrez* search and retrieval system, familiar to *PubMed* users, which provides seamless access to over 40 molecular and literature databases including *PubMed*.

OMIM provides background information on diseases with hyperlinks to related journal articles in *PubMed*. Students need to select the correct entry or record from the *OMIM* results. All entries in the *OMIM* database are preceded by symbols designating the molecular status of each entry. An asterisk (*) denotes data on a gene of a known genetic sequence, a pound sign (#) or plus sign (+) indicate a descriptive entry with no known genetic locus or known sequence. Other symbols usually indicate an ambiguous genetic origin with less relevant or minimal data, so students are advised to select an *OMIM* entry or record preceded by an asterisk or pound symbol if necessary as they would provide more information. Students who select various cancers or Alzheimer's disease which are characterized by a complex etiology, numerous causative genes or mutations and possible environmental factors, often have to check or corroborate their findings with related *PubMed* articles or other data sources used in this exercise to ensure that they have identified the most relevant *OMIM* entry. Each *OMIM*record is well organized and includes a *Table of Contents* comprised of approximately twenty sections such as *Description, Gene Structure, Mapping* and*Biochemical features* as well as an *External Links* section with links to other data sources covering *DNA, Genome, Gene Info, Protein* and *Clinical Resources.* Students then use the *Gene Info* link on the *OMIM* record as a gateway to *NCBI's Gene*database to locate data pertaining to the gene's architecture and function for the next part of their assignment.

NCBI *Gene*

(http://www.ncbi.nlm.nih.gov/gene/)

Gene provides extensive gene-specific data including gene structure or architecture such as mRNA and protein length. Students use the *Gene* database to better understand gene structure and the effect mutations have on related functions. Like *OMIM, Gene* uses *NCBI's Entrez* discovery tool and includes a *Table of Contents* and *Related Information* for convenient access to the data students needed to answer questions. Questions included the length of the wild-type protein which could be found using the "*RefSeqProteins*" link, the length or size of the mature mRNA (e.g., *RefSeqRNAs*) for their gene, and the number of exons and introns (e.g., *RefSeqGene*). Exons are nucleotide sequences that code for proteins while introns do not. Students also identify the type of chromosome where the gene is located (e.g., metacentric, telocentric, or acrocentric), and locate an ideogram or schematic diagram (e.g., *Map Viewer*) displaying this information. All of this data is available from the *Related Information* section on the *Gene* page.

PubMed

(http://www.ncbi.nlm.nih.gov/pubmed)

Each student has to obtain one review article on their genetic disease topic to provide background information and supplement and contextualize the data in their poster presentations.

Google Patents

(https://www.google.com/?tbm=pts&gws_rd=ssl)

Patents provide a unique source of scientific information particularly in the life sciences where many researchers often seek patent protection to recover research and development costs associated with research conducted on genes, proteins, laboratory techniques and related diagnostic tools. Patents in these fields frequently contain pertinent molecular information including images that help students with their research. Students locate one patent that relates to the molecular origins of their disease topic such as the causative gene or protein, or a diagnostic method used to detect the mutant gene or protein. Patents that relate to therapies or treatments for their disease topic are not used for this assignment.

As this is likely the students' first introduction to patents and they only need to locate one, *Google Patents* is used to access patents from the *United States Patent and Trademark Office*. *Google Patents* provides a user-friendly interface for searching patents from the major patent issuing agencies. Students answer questions regarding the patent number, filing date, and details about the patent's "Claims" as well as a discussion question regarding the ethics of patenting life forms such as genes and patents and whether they agree that their chosen patent should have been granted.

By completing the lab workbook, students gather much of the information they need for their poster presentation. Students had several weeks to develop the poster, during which time they can refer to the lab workbook and LibGuide, or consult with the librarian.

4.2 BCEM 393-Introduction to Biochemistry

This course, like Biology 311 which is a pre-requisite, is required for all biology majors, and is offered in the winter semester. Whereas the Biology 311 information literacy workshop focuses on the molecular basis of genetically-inherited diseases, the library lab in Biochemistry 393 concentrates on developing a structural understanding of the molecular basis of these diseases, particularly the effects of mutations on protein structure and function. This course covers the structure and function of carbohydrates, amino acids, proteins, lipids, coenzymes and enzymes including an introduction to metabolism and energy transduction. The information literacy workshop is integrated into Lab 2, "Gaining a structural perspective on the molecular basis of inherited diseases", and is usually held during the first week of February.

The teaching objectives for this lab include:

- Use the *Protein Data Bank* to locate the protein's three-dimensional coordinates.
- Use *ProtParam* to calculate the isolectric point of a protein.
- Run a Blast search to identify homologous or similar proteins and use *ClustalW*to align them.
- Use *PyMOL to* investigate the effect of a mutation on protein structure and function.

Students select a human genetically-inherited disease topic from a list of eleven approved disease topics chosen because they have associated files and designated mutations that would work with the appropriate bioinformatics resources in this assignment. In order to focus students' attention and effort on using the data, the instructor does more advance work, including identifying appropriate topics and articles, and pre-loading corresponding *PyMOL* files on the course website. Like the work students complete for their Biology 311 project, students in this class are introduced to several bioinformatics databases that serve as data repositories for protein structural information where they examine the impact of mutations on protein structure, function and specific diseases.

Students begin their research with a 'seed' or primary article pre-selected by the instructor to provide background and structural information on their chosen protein without discussing the mutation they will be studying. Students need to develop and test a hypothesis about the mutation based on the data they collect from the various bioinformatics sources in order to complete their assignment and presentation.

The librarian leads the students through the exercises using the "K-Ras" protein as an example, and screen shots are included with accompanying text in the lab workbook. Students start their research with familiar resources introduced in Biology 311, such as *Genetics Home Reference*, *PubMed* and *OMIM* to locate background information about their disease including etiology and enzyme(s). They then explore specific protein data repositories to retrieve information about their chosen protein.

These data resources are highly integrated with each other, enabling students to link seamlessly between each source to gather required information. For example, students using the *Protein Data Bank* (PDB) to obtain structural information about their chosen protein can also retrieve the '*Primary Citation*'article associated with that protein by using the *PubMed ID* or *Digital Object Identifier* (DOI) which is hyperlinked to *PubMed* and/or the publisher's site. Students obtain information about the size of the protein (e.g., number of amino acids) associated with their chosen topic. The size or length of the protein can vary slightly among different bioinformatics databases but students are shown how to check the actual protein size by selecting the *UniProtKB* number hyperlink which automatically directs them to *ProtParam,* another protein structural repository, for the most current information. The *ProtParam* protein repository is subsequently used to locate protein sequence data.

Protein Data Bank (PDB)

(http://www.rcsb.org/pdb/home/home.do)

The *Protein Data Bank (PDB)* is a portal for three-dimensional structural information on over 100,000 proteins and nucleic acids. A major component of the project for this lab was having students locate the 3D coordinates for the protein associated with their genetically-inherited disease topic. Because a protein's function is intricately linked to its structure, any change or degradation in a protein's three-dimensional structure will result in a loss of protein function. Consequently, protein structural information is critical to understanding a protein's function. For this portion of the lab students are guided through a series of questions on their selected protein's structure, including protein size, molecular weight, and number of amino acids. Each *PDB*record also has a unique *UniProtKB* reference link that redirects the user from the*PDB* record to the corresponding *UniProtKB* entry, where students are able to locate more structural information about their protein and perform a *BLAST*search to locate similar proteins to their selected protein.

UniProtKB

(http://www.uniprot.org/)

UniProtKB is a clearinghouse for sequence and functional information about proteins. Students use *UniProtKB* it to find data including the molecular weight, isoelectric point and number of amino acids for their protein.

BLAST

(http://www.uniprot.org/blast/)

Students performed a *BLAST* search to locate any similar or homologous proteins and measure the degree of similarity between different protein sequences. This enables students to determine what function a certain protein fulfills among various organisms. Students use *BLAST* to determine if there are any protein sequences from *Gallus gallus* (Chicken), *Mus musculus* (Mouse), or *Danio rerio* (Zebrafish) similar to their chosen protein. By examining their results students deduce what function their chosen protein might perform. Students also conduct a sequence alignment between these four organisms using the *ClustalW* program which is embedded within the *BLAST* search results. They can then determine the best match between the selected sequences.

PyMOL

(http://www.pymol.org/)

PyMOL is an open source molecular visualization program that enables users to manipulate and view protein structures in three-dimensions. Students used*PyMOL* to study the effects of a single amino acid substitution on protein structure and function. *PyMOL* is downloaded in advance on the classroom workstations and each student has access to a *PyMOL* file consisting of a pre-selected view of each protein structure through their course management software pack. After uploading their specific *PyMOL* file students can manipulate and rotate their file using different function keys in order to view their protein structure from different perspectives. Students then insert their assigned disease-causing mutation in the protein by substituting an amino acid. For demonstration purposes the librarian uses the "K-RAS" protein to demonstrate mutation *G60R*, where *Gly-60* (Glycine) was substituted by *Arg-60* (Arginine). Using their own protein examples, students complete a series of exercises that investigate the impact of mutations on the three-dimensional structure of their protein. Their observations and collected data are incorporated in the poster presentation.

Cologne University Protein Stability Analysis Tool (CUPSAT)

(http://cupsat.tu-bs.de/)

The final data source presented in this session is *CUPSAT,* which enables students to predict changes in protein stability when point mutations are introduced. This type of mutation changes a single nucleotide base pair. Students carry out a number of tasks and include the results in their research.

Students work in pairs during this workshop and while most students complete the assigned questions during the session, they have approximately 5 weeks to finish it and prepare their poster for the oral presentation at the end of term. Students also complete a series of independent exercises at regular intervals throughout the term which integrate course content into the work they do on the poster presentation.

5. RESULTS

Incorporating disciplinary data from gene and protein data sources within undergraduate IL has benefits for librarians and students. For the librarian, developing familiarity with life science data provided an opportunity to engage with the discipline more deeply. Developing competencies through the discovery and analysis of molecular biology data has informed the librarian's domain expertise in genetics and biochemistry, in turn providing valuable contextual knowledge when demonstrating and answering questions about the various data resources. Gaining knowledge of bioinformatics resources has also translated into an improved alignment of library services with faculty requirements, and has resulted in many productive conversations with faculty members about their data use and needs. This in turn has led to a more collaborative, sustainable and rigorous information literacy program. The value of this collaboration with a faculty members extends beyond the classroom, and has resulted in numerous presentations to both library and biology audiences.

As mentioned earlier in this paper, librarians, particularly science librarians, need to maintain currency in patterns of scholarly communication, to better align their professional activities more closely with the research and learning needs of students and faculty members. Jaguszewski and Williams posit that "an engaged librarian seeks to enhance scholar productivity, to empower learners, and to participate in the entire lifecycle of the research, teaching, and learning process" (**Jaguszewski & Williams, 2013**, p. 4). This need to develop new skills in order to take on new roles is a dominant theme in current literature (**Auckland, 2012**; **Kenney, 2014**).

Students make significant learning gains through the experiential learning provided in the library sessions. Using bioinformatics databases to answer specific molecular biology questions ensured that students were able to extend and integrate learning throughout the academic year, progressing from easy to more complex questions, while at the same time understanding concepts that underline both disciplines. In both courses, bioinformatics tools facilitated the discovery and analysis of life sciences data, illustrating the role of gene and protein structure in determining related functions. Students can learn about these topics in a lecture or from a textbook but using *PyMol* to substitute an amino acid and visualize the impact on protein structure and function in three-dimensions provides a more engaged, interactive and authentic learning experience. Using these databases in conjunction with other sources such as *PDB*and *PubMed* allowed for a more integrated understanding of how various sources of information serve researchers. Using a *GenBank* record, a patent, and a scholarly article that refer to the same data demonstrates how these resources complement each other, and how students can correlate information found across the resources to better understand it. Starting from genetics resources to solve problems in biochemistry underscores the deep connections between the two disciplines and encourages deeper transfer of knowledge.

In considering the gains students make through this integration beyond the simple exposure to new resources, it is possible to relate them to the threshold concepts in information literacy currently under discussion by the ACRL. Many of the Knowledge Practices described in the Framework document are supported by the design of the library session and accompanying assignments (**Gibson & Jacobson, 2014**) . Students are encouraged to develop a deeper understanding of how knowledge in the discipline is created and deploy that understanding to search and research more effectively, in order to synthesize information from a variety of sources into a scholarly artefact. Students develop an understanding of bioinformatics data in BIOL 311 through the library session and subsequent practice. They then deepen that understanding in BCEM 393, through exercises deliberately designed to foster knowledge transfer, and again have multiple opportunities to practice extracting information from data.

Students who participated in these sessions gained hands-on experience with critical tools in their discipline, developed an appreciation for how knowledge in genetics and biochemistry is created and integrated based on incremental discoveries, and how new bioinformatics tools facilitate the synthesis of information into knowledge. Their instructor has noted that students have developed a more integrated understanding of biochemistry and genetics, in part through seeing how researchers use similar data to understand questions in both domains. The librarian frequently attends the poster sessions and has observed students demonstrate an ability to use the bioinformatics data in conjunction with more conventional text resources to explore genetic factors in disease. Some of the Graduate Teaching Assistants, now working with the classes, were early participants in the library sessions and frequently comment on the usefulness of early exposure to the databases.

Throughout this collaboration, assignments and activities have been continuously revised in response to new resources, new interfaces and observations of where students have been challenged to understand the material or the processes to access and link material. From experience and the literature, the following suggestions for incorporating bioinformatics data emerge. Many of these are common to IL as a whole, but worth reconsidering and relating to new content:

- Be prepared to put in substantial time developing familiarity with the bioinformatics databases and subject content in order to develop exercises and demonstrations and to anticipate problems and student questions.
- Be prepared to revise constantly as the tools change frequently.
- Instruction in bioinformatics data should be course integrated, with a graded assignment of sufficient weight so that students take it seriously.
- Develop sequential steps from simple databases and operations to more complex ones.
- Develop specific questions for each resource that highlight the relevant data in each.
- Scaffold exercises to integrate just-learned material with new steps.
- Explicitly link the content to the course, and the content of different resources to each other to help students develop a structure for the new knowledge.
- Students learn by doing, allow hands-on interactivity & flexibility, and multiple opportunities to practice.
- Consider integrating text and data resources so students see how the same information is used in different ways.

6. CONCLUSION

This case study presents the development of two linked IL sessions that introduce undergraduate students to bioinformatics resources. While each of the classes provides benefits to students in terms of enhanced content knowledge, linking the courses through their foundational information resources helps students understand genetics and biochemistry as integrated fields of inquiry. This has clear benefits for the students, and indeed for the librarian. Moving beyond the consideration of IL as isolated experiences, to a more integrated curriculum allows for a deeper engagement with student learning and a closer alignment with the needs of current and future researchers. This closer alignment has led to more effective collaborations, not only with the instructor in these courses, but also with other members of the department. Developing an understanding of the role of bioinformatics databases has allowed the librarian to work with current researchers at a different level, as well as providing future researchers with necessary skills.

REFERENCES

1. Akers, K.G., & Doty, J. (2013). Disciplinary differences in faculty research data management practices and perspectives. International Journal of Digital Curation, 8(2), 5–26.

2. Auckland, M. (2012). Re-skilling for research: An investigation into the role and skills of subject and liaison librarians required to effectively support the evolving information needs of researchers (p. 115). London: Research Libraries UK. Retrieved January 7, 2015, from http://www.rluk.ac.uk/wp-content/uploads/2014/02/RLUK-Re-skilling.pdf

3. Badotti, F., Barbosa, A.S., Reis, A.L.M., do Valle, I.F., Ambrósio, L., & Bitar, M. (2013). Comparative modeling of proteins: A method for engaging students interest in bioinformatics tools. Biochemistry and

Molecular Biology Education: A Bimonthly Publication of the International Union of Biochemistry and Molecular Biology, 42(1), 68–78.

4. Barrette-Ng, I., & MacMillan, D. (2014). Gaining a structural perspective on the molecular basis of inherited diseases. In Tested Studies for Laboratory Teaching: Proceedings of the 35th Workshop/Conference of the Association for Biology Laboratory Education (ABLE) (pp. 225–244). Calgary, Alberta: Association for Biology Laboratory Education
5. Bednarski, A.E., Elgin, S.C.R., & Pakrasi, H.B. (2005). An inquiry into protein structure and genetic disease: Introducing undergraduates to bioinformatics in a large introductory course. Cell Biology Education, 4(3), 207–220. doi: 10.1187/cbe.04–07–0044
6. Bourne, P. (2005). Will a biological database be different from a biological journal? PLoS Computational Biology, 1(3), 179–181.
7. Bowden, T.S., & DiBenedetto, A. (2001). Information literacy in a biology laboratory session. Research Strategies, 18(2), 143–149.
8. Boyle, J.A. (2004). Bioinformatics in undergraduate education. Biochemistry and Molecular Biology Education, 32(4), 236–238. Retrieved January 7, 2015
9. Brown, C., & Krumholz, L.R. (2002). Integrating information literacy into the science curriculum. College & Research Libraries, 63(2), 111–123.
10. Callinan, J.E. (2005). Information-seeking behaviour of undergraduate biology students. Library Review, 54(2), 86–99.
11. Calzada Prado, J., & Marzal, M.Á. (2013). Incorporating data literacy into information literacy programs: Core competencies and concepts. Libri: International Journal of Libraries & Information Services, 63(2), 123–134.
12. Carlson, J., Johnston, L., Westra, B., & Nichols, M. (2013). Developing an approach for data management education: A report

from the data information literacy project. International Journal of Digital Curation, 8(1), 204–217.

13. Castelli, D., Manghi, P., & Thanos, C. (2013). A vision towards scientific communication infrastructures. International Journal on Digital Libraries, 13(3–4), 155–169.
14. Dinkelman, A.L. (2010). Using course syllabi to access research expectations of biology majors. Issues in Science & Technology Libraries, Winter.
15. Ditty, J.L., Kvaal, C.A., Goodner, B., Freyermuth, S.K., Bailey, C., Britton, R.A., Kerfeld, C.A. (2010). Incorporating genomics and bioinformatics across the life sciences curriculum. PLoS Biology, 8(8), e1000448.
16. Dymond, J.S., Scheifele, L.Z., Richardson, S., Lee, P., Chandrasegaran, S., Bader, J.S., & Boeke, J.D. (2009). Teaching synthetic biology, bioinformatics and engineering to undergraduates: The interdisciplinary Build-a-Genome course. Genetics, 181(1), 13–21.
17. Ferrer-Vinent, I.J., & Carello, C.A. (2008). Embedded library instruction in a first-year biology laboratory course. Science & Technology Libraries, 28(4), 325–351.
18. Gibson, C., & Jacobson, T.E. (2014). ACRL Framework for information literacy for higher education (p. 26). Chicago, IL: Association of College & Research Libraries. Retrieved January 7, 2015, from http://acrl.ala.org/ilstandards/wp-content/uploads/2014/11/Framework-for-IL-for-HE-draft-3.pdf
19. Gray, A.S. (2004). Data and statistical literacy for librarians. IASSIST Quarterly, 28(2/3), 24–29. Retrieved from http:// www. iassistdata. org/sites/ default/files/ iq/iqvol282_ 3gray. pdf
20. Hey, T., Tansley, S., & Tolle, K. (2009). Jim Gray's fourth paradigm and the construction of the scientific record (pp. 177–183). Redmond, WA: Microsoft Corporation. Retrieved January 7, 2015, from http://research.microsoft.com/en-us/collaboration/ fourth paradigm/4th_paradigm_book_complete_lr.pdf

21. Jaguszewski, J.M., & Williams, K. (2013). New roles for new times: Transforming liaison roles in research libraries. Washington DC: Association of Research Libraries (p. 17). Retrieved January 7, 2015, from http://www.arl.org/storage/documents/publications/NRNT-Liaison-Roles-final.pdf

22. Kenney, A.R. (2014). Leveraging the liaison model: From defining 21st century research libraries to implementing 21st century research universities. Ithaka (p. 11). Retrieved January 7, 2015, from http://www.sr.ithaka.org/blog-individual/leveraging-liaison-model-defining-21st-century-research-libraries-implementing-21st

23. MacMillan, D. (2010). Sequencing genetics information: Integrating data into information literacy for undergraduate biology students. Issues in Science & Technology Libraries, Spring.

24. MacMullen, W.J., & Denn, S.O. (2005). Information problems in molecular biology and bioinformatics. Journal of the American Society for Information Science and Technology, 56(5), 447–456.

25. Miskowski, J.A., Howard, D.R., Abler, M.L., & Grunwald, S.K. (2007). Design and implementation of an interdepartmental bioinformatics program across life science curricula. Biochemistry and Molecular Biology Education: A Bimonthly Publication of the International Union of Biochemistry and Molecular Biology, 35(1), 9–15.

26. Schield, M. (2004). Information literacy, statistical literacy and data literacy. IASSIST Quarterly, 28(2/3), 6–11. Retrieved January 7, 2015, from http://www.statlit.org/pdf/2005SchieldIASSIST.pdf

27. Smith, A., Balazinska, M., Baru, C., Gomelsky, M., McLennan, M., Rose, L., Kolker, E. (2011). Biology and data-intensive scientific discovery in the beginning of the 21st century. Omics: A Journal of Integrative Biology, 15(4), 209–212.

28. Stephenson, E., & Schifter Caravello, P. (2007). Incorporating data literacy into undergraduate information literacy programs in the social sciences. Reference Services Review, 35(4), 525–540.

29. Via, A., De Las Rivas, J., Attwood, T.K., Landsman, D., Brazas, M.D., Leunissen, J.A.M., Schneider, M.V. (2011). Ten simple rules for

developing a short bioinformatics training course. PLoS Computational Biology, 7(10), e1002245.

30. Vickers, A.J. (2011). Making raw data more widely available. BMJ (Clinical Research Ed.), 342(May04_2), d2323.

31. Voet, J.G., Bell, E., Boyer, R., Boyle, J., O'Leary, M., & Zimmerman, J.K. (2003). Recommended curriculum for a program in biochemistry and molecular biology. Biochemistry and Molecular Biology Education, 31(3), 161–162.

32. Weisman, D. (2010). Incorporating a collaborative web-based virtual laboratory in an undergraduate bioinformatics course. Biochemistry and Molecular Biology Education: A Bimonthly Publication of the International Union of Biochemistry and Molecular Biology, 38(1), 4–9.

33. Wightman, B., & Hark, A.T. (2012). Integration of bioinformatics into an undergraduate biology curriculum and the impact on development of mathematical skills. Biochemistry and Molecular Biology Education: A Bimonthly Publication of the International Union of Biochemistry and Molecular Biology, 40(5), 310–319.

34. Winterman, B. (2009). Building better biology undergraduates through information literacy integration. Issues in Science & Technology Libraries, Summer.

Chapter 12

Evaluation as a Tool for Developing the Quality of the Academic Libraries. Case Study at a Large and Traditional Research Oriented Scandinavian University - Viewpoint of a Member of the Evaluation Panel

Gunnar Sahlin

ABSTRACT

We were invited as an international review team to assess the development of the library system in the Helsinki University. As reviewers and experts we came to Helsinki for two days in May 2004 - Hans Geleijnse from The Netherlands, Sinikka Koskiala from Finland and myself from Sweden. The result of our evaluation is the report Follow-up Evaluation of Library and information Services in Helsinki University 2004 (Geleijnse, Koskiala & Sahlin, 2004).

1. THE COMMISSION

We were invited as an international review team to assess the development of the library system in the Helsinki University. As reviewers and experts we came to Helsinki for two days in May 2004 - Hans Geleijnse from The Netherlands, Sinikka Koskiala from Finland and myself from Sweden. The result of our evaluation is the report Follow-up Evaluation of Library and information Services in Helsinki University 2004 (Geleijnse, Koskiala & Sahlin, 2004).

The university requested the panel to answer some questions, and they wanted us to give some comments on the key issues for the future. The questions and key issues were the starting point for our investigation. Another base for our study was the assessment of the library activities in the university from the year 2000 (Geleijnse et al, 2000). The strategy for the university, the assessment report with different viewpoints and recommendations and other relevant information were useful for us in the evaluation. How has the university succeeded in its library development, required by the panel from year 2000?

2. THE EVALUATION PROCESS

The evaluation started the year before with self-evaluations of librarians and library staff, deans, directors of independent institutes, researchers, teachers, and students. The reports from these self-evaluations were provided to us. Besides the report from year 2000 and the self-evaluations we took part of other written material, from the libraries and the university, before our visit to Helsinki. The strategy for the university, action plans for libraries and the information technology, data and statistics concerning library activities, annual reports and other relevant information were delivered to us. All this material was very useful in the evaluation.

A steering-group, chaired by Vice Rector and chairman of the Library Committee, Hannele Niemi, was formed in 2003 to prepare and co-ordinate the review. They, together with the evaluation officer and the Director of Information and Library Services, gave us good support. Besides the steering group we met the Rector and deans, groups of researchers, teachers, students, and of course library staff. The National Librarian and Directors for the faculty and campus libraries gave us their descriptions. We visited branch libraries and the educations centre, Alexandria Learning Centre.

The written material, the visits, and the group-meetings - all gave us a broad and clear view of the library situation at the University of Helsinki. Following the visit our notes and impressions blended together - in close contacts through e-mail - to the evaluation report.

Even if the evaluation is two years old, it's no problem to remember the meetings, the visits and the results. First of all I would like to express that I'm satisfied with the evaluation - the process and the report. It gave me, and I hope also the other in the evaluation group, very interesting impressions, which are useful in several of following situations. When you are part of an evaluation group you give your knowledge and your impressions. At the same time you get something back, new experiences which are useful for yourself.

The second thing I will say is about decentralization. I have worked in several universities - as researcher, teacher, library director and as member of board - and every university is by nature a decentralized organization. But Helsinki University is one of the most decentralized universities I have seen. It was important to notice this fact when we studied the situation for the library organization.

3. THE EVALUATION REPORT

In the panel we were pleased with the activities following the assessment year 2000. One example is the creation of a small central unit, responsible for cooperation and electronic resources, and it's important to have this unit permanent.

Despite the development in the university libraries from year 2000, we identified during our evaluation process some important obstacles, problems or questions. I will give you some examples of these and the decentralization, funding model, and the role of the National Library are the common denominators for these obstacles, problems or questions. A special situation for Finland was the National library's close relation to the university. They still get their financial support through the university but the situation has improved. For the other problems we scrutinized many library directors who can find some relations to their own situation:

- a continuous tension concerning what should be done on a central level and what on a decentralized level in the campus libraries,
- the management of the important differences between the campus libraries,
- the need for a solid funding model for the future,
- a coherent staff strategy,
- the cost of space and the fact the available space is partly underused,
- the need to make better use of the expensive information resources,
- clear service agreements between the University of Helsinki and the National Library.

In our study of the structure and organization of the university libraries we notice the important and strong role of the campus libraries, the new role of the National Library and the importance of the central Unit of Information and Library Services and Development. Together they form what we call 'The University Library'. In our recommendations we noticed - some examples - a need

- for further cooperation between these three parts but also between the different campus libraries,
- for a sustainable staff strategy to better use of the human resources,
- further clarifications of the relations between the university and the National Library, with tasks being performed by the National Library and service level agreements were needed,
- to increase the funding for licensing and electronic information.

Some of our ideas in the evaluation group for the development in the university and in the National Library have been fulfilled. Of course it depends on great activities from those persons responsible for the development, but for us in the evaluation group it's a pleasure to have been a small part of this development in the library of Helsinki University.

REFERENCES

1. Geleijnse, Hans, Göran Gellerstam, Ian R.M. Mowat and Kyllikki Ruokonen; introduction by Aimo Virtanen: Helsinki University Libraries - Report of an Assessment panel. Helsinki: University of Helsinki, 2000. (Evaluation Projects of the University of Helsinki, 7).

2. Geleijnse, Hans, Sinikka Koskiala and Gunnar Sahlin: Follow-up Evaluation of the Library and Information Services 2004. Helsinki, 2004. (Evalution Projects of the University of Helsinki, 8).

Index